AF361768

Graphene Nanoplatelets

Graphene Nanoplatelets

Special Issue Editors

Silvia González Prolongo
Alberto Jiménez Suárez

MDPI • Basel • Beijing • Wuhan • Barcelona • Belgrade

Special Issue Editors

Silvia González Prolongo
Department of Applied
Mathematics, Materials Science
and Engineering and Electronic
Technology, Universidad Rey
Juan Carlos
Spain

Alberto Jiménez Suárez
Department of Applied
Mathematics, Materials Science
and Engineering and Electronic
Technology, Universidad Rey
Juan Carlos
Spain

Editorial Office
MDPI
St. Alban-Anlage 66
4052 Basel, Switzerland

This is a reprint of articles from the Special Issue published online in the open access journal *Applied Sciences* (ISSN 2076-3417) from 2018 to 2020 (available at: https://www.mdpi.com/journal/applsci/special_issues/Graphene_Nanoplatelets).

For citation purposes, cite each article independently as indicated on the article page online and as indicated below:

LastName, A.A.; LastName, B.B.; LastName, C.C. Article Title. *Journal Name* **Year**, *Article Number*, Page Range.

ISBN 978-3-03928-794-9 (Pbk)
ISBN 978-3-03928-795-6 (PDF)

Cover image courtesy of Gilberto Del Rosario Hernández.

Contents

About the Special Issue Editors

Silvia González Prolongo is Full Professor in Materials Engineering at the University Rey Juan Carlos (Madrid, Spain). She is the head of the Smart Nanocomposite and Polymers research group. Her main research lines are aimed at processing and optimizing of smart multifunctional composites, and developing new functionalities, such as self-sensors, self-actuators, self-heaters, and self-healable materials;, for aerospace, solar and wind energy, automotive, and packing industries, amongst others.

Albert Jiménez-Suárez, Industrial Engineer, Ph.D. in Materials Science and Engineering, and Senior Lecturer at Universidad Rey Juan Carlos (Spain). His research work focuses on processing and characterizing of multifunctional composites based on the addition of nanoreinforcements to polymer matrices, including their use as adhesives, coatings, matrices for multiscale reinforced composites, and 3D printing technologies.

 *applied sciences*

Editorial

Graphene Nanoplatelets

A. Jiménez-Suárez and S. G. Prolongo *

Area of Materials Science and Engineering. ESCET. University Rey Juan Carlos, c/Tulipán s/n, Móstoles, 28933 Madrid, Spain; alberto.jimenez.suarez@urjc.es
* Correspondence: silvia.gonzalez@urjc.es; Tel.: +34-91488-82-92

Received: 17 February 2020; Accepted: 25 February 2020; Published: 4 March 2020

Featured Application: The excellent performance of graphene nanoplatelets turns them into engaging fillers for different materials, offering a wide range of applications from energy harvesting, flexible electronic devices, smart sensors and structural-functional composites.

---

1. Structure, Morphology, Properties and Behavior

Graphene is regarded as the revolutionary material of the 21st century. It is a single graphite monolayer, whose thickness is one atom (0.34 nm) while its lateral size could be several orders of magnitude larger. Its synthesis is complex and cannot be mass-produced yet. For this reason, graphene nanoplatelets (GNPs) have become an alternative, with a low cost and exciting properties, and the potential for large-scale production. GNPs have few graphite layers, varying in thickness from 0.7 to 100 nm [1].

Their main properties are light weight, high aspect ratio with planar shape, good mechanical properties and excellent thermal and electrical conductivities, together with low cost and easy manufacture. GNPs have numerous applications as isolated materials, neat coatings and fillers of composites. This Special Issue is focused on the use of graphene nanoplatelets as nanofillers [2–4].

2. Current and Future Applications of GNP Nanocomposites

Graphene nanoplatelets are widely employed as nanofillers with different matrices, such as polymers, concretes, metals, among others. The addition of GNP usually enhances the mechanical and tribological behavior, increasing the barrier properties and thermal conductivity, transforming insulating matrices into electrical conductors and acting as a flame retardant.

The manufacture of these nanocomposites is a challenging task to get a suitable GNP dispersion. The added GNP content varies significantly as a function of the nature of the matrix and the required properties. Due to their great versatility, the reasons and expectations raised by graphene nanoplatelets addition are very varied, looking for different performances and therefore applications. The current and future applications of these nanocomposites are unlimited, from materials with enhanced mechanical and thermal behavior up to new functional materials, such as sensors, new electronic devices, energy harvesting, adsorbents, etc. Several specific examples of the wide versatility of nanocomposites reinforced with graphene nanoplatelets are collected in the different works of this Special Issue and they are summarized hereunder.

As was just mentioned, the dispersion of graphene nanoplatelets together with the achieved exfoliation degree affects the properties and behavior of manufactured composites. For this reason, their study is essential in the development of these materials. Terahertz time-domain spectroscopy (THz-TDS) [5] is a new technique to provide information regarding graphene dispersion, analyzing the dielectric behavior of the material. Also, it enables investigating in situ the electronic quality of the polymer nanocomposite.

GNPs are commonly added to polymer matrices to enhance their mechanical behavior, increasing their chemical resistance and therefore their lifetime [6]. GNPs added into polymers reduce their ability to absorb water, increasing their resistance to aggressive humid environments. The barrier properties of GNPs are associated with the formation of tortuous paths for the water molecules, depending markedly on their geometry, thickness and lateral size. This is due to their spatial arrangement which modifies the effective specific surface area of GNPs.

The composites with polyethylene glycol (PGE) matrix present high performance in photothermal energy conversion, showing higher absorption efficiency for solar irradiation [7]. This, together with the enhanced thermal conductivity by GNP addition, makes these nanocomposites in promising materials for solar energy conversion and storage.

GNPs are also added to concrete in order to improve freeze–thaw (F–T) resistance [8]. Concretes reinforced with GNPs have enhanced compressive strength and F–T durability due to their finer pore structure than ordinary concretes. Due to this habitual tendency, there is a specific amount of graphene nanoplatelets which provides the best behavior.

Graphene oxide (GO) nanoplatelets are added to sandwich composites, such as Fe-based metal organic frameworks [9], to enhance their efficiency in the capture of uranium. GO improves the absorption of hexavalent uranium. This is an important advancement for using nuclear energy in a safe way, helping the elimination of water U-contamination. Advanced composite absorbents based on GO have been manufactured with satisfactory radionuclide uptake ability in regard to individual components.

The application of graphene nanoplatelets is very varied, with them being adding to solids, semi-liquids or greases, and also liquids. An example of the reinforcement of liquids is their addition to oils or grease lubricants and water.

Water-based graphene composite [10] presents enhanced thermal conductivity, while its freezing and melting time decrease with the graphene volume added. This behavior of water is interesting for storing electrical energy in batteries or as compressed air storage.

On the other hand, sometimes, the individual GNP addition is not enough, requiring the combination with other fillers to obtain the required behavior. In fact, in these cases, a synergic effect between both nanofillers is looked for. This is the case of the GNP addition in liquid lubricants [11]. The combined use of two additives, GNP and titanium dioxide nanopowders, enhances its wear and friction properties. It is worthy to note that graphene nanoplatelets are enough of a filler to reduce the friction of grease lubricant, without any additive.

All these cases confirm the versatility of graphene nanoplatelets as nanofillers of very different materials, developing nanocomposites with enhanced behavior and new functionalities, which can be used in very different applications. For this reason, graphene nanoplatelets are considered one of the most outstanding nanofillers in the last decade, which will be able to bring about a revolution in society, providing it with new devices and developments.

Acknowledgments: The authors acknowledge the Ministerio de Economía y Competitividad of Spain Government (MAT2016-78825-C2-1-R) and Comunidad de Madrid Government (ADITIMAT-CM C2018/NMT-4411).

Conflicts of Interest: The authors declare no conflicts of interest.

References

1. Choi, W.; Lahiri, I.; Seelaboyina, R.; Kang, Y.S. Synthesis of graphene ad its applications: A review. *Crict. Rev. Sol. Stat. Mater. Sci.* **2010**, *35*, 52–71. [CrossRef]
2. Potts, J.R.; Dreyer, R.D.; Bielwski, C.W.; Ruoff, R.S. Graphene-based polymer nanocomposites. *Polymer* **2011**, *52*, 5–25. [CrossRef]
3. Lawal, A.T. Graphene-based nanocomposites and their applications. A review. *Biosens. Biolectron.* **2019**, *141*, 111384. [CrossRef] [PubMed]
4. Navasingh, R.J.H.; Kumar, R.; Marimuthu, K.; Planichamy, S.; Khan, A.; Asiri, A.M.; Asad, M. Graphene-based nano metal matrix composites: A review. In *Nanocarbon and Its Composites*; Series in Composites Science and Engineering; Elsevier Sci LTD: London, UK; Woodhead Publishing: Cambridge, UK, 2019; pp. 153–170.

Appl. Sci. **2020**, *10*, 1753

5.	Cui, H.; Zhang, X.B.; Yang, P.; Su, J.F.; Wei, X.Y.; Guo, Y.H. Spectral characteristic of single layer graphene via terahertz time domain spectroscopy. *Optik* **2015**, *126*, 1362–1365. [CrossRef]
6.	Arribas, C.; Prolongo, M.G.; Sánchez-Cabezudo, M.; Moriche, R.; Prolongo, S.G. Hydrothermal ageing of graphene/carbon nanotubes/epoxy hybrid nanocomposites. *Polym. Degrad. Stab.* **2019**, *170*, 109003. [CrossRef]
7.	Wang, F.; Zhang, P.; Mou, Y.; Kang, Y.; Liu, M.; Song, L.; Lu, A.; Rong, J. Synthesis of the polyethylene glycol solid-solid phase change materials with a functionalized graphene oxide for thermal energy storage. *Polym. Test.* **2017**, *63*, 494–504. [CrossRef]
8.	Shamsaei, E.; Souza, B.; yao Xm Benhelal, E.; Akbari, A.; Duan, W. Graphene-based nanosheets for stronger and more durable concrete: A review. *Const. Build. Mater.* **2018**, *183*, 642–660. [CrossRef]
9.	Jiahui, Z.; Hongsen, Z.; Qi, L.; Cheng, W.; Zhiyao, S.; Rumin, L.; Peili, L.; Milin, Z.; Jun, W. Metal-organic frameworks (MIL-68) decorated graphene oxide for highly efficient enrichment of uranium. *J. Taiwan Inst. Chem. Eng.* **2019**, *99*, 45–52.
10.	Foroutan, M.; Fatemi, S.M.; Shokouh, F. Graphene confinement effects on melting/freezing point and structure and dynamics behavior of water. *J. Molec. Grap. Model.* **2016**, *66*, 85–90. [CrossRef] [PubMed]
11.	Penkov, O.V. Graphene-based lubricants. In *Tribology of Graphene*; Elsevier: Amsterdam, The Netherlands, 2020; pp. 193–236. ISBN 978-0-12-818641-1.

 *applied sciences*

Review

Graphene Nanoplatelets-Based Advanced Materials and Recent Progress in Sustainable Applications

Pietro Cataldi *, Athanassia Athanassiou and Ilker S. Bayer

Smart Materials, Istituto Italiano di Tecnologia, Via Morego 30, 16163 Genova, Italy;
athanassia.athanassiou@iit.it (A.A.); ilker.bayer@iit.it (I.S.B.)
* Correspondence: pietro.cataldi@iit.it

Received: 5 July 2018; Accepted: 15 August 2018; Published: 23 August 2018

Abstract: Graphene is the first 2D crystal ever isolated by mankind. It consists of a single graphite layer, and its exceptional properties are revolutionizing material science. However, there is still a lack of convenient mass-production methods to obtain defect-free monolayer graphene. In contrast, graphene nanoplatelets, hybrids between graphene and graphite, are already industrially available. Such nanomaterials are attractive, considering their planar structure, light weight, high aspect ratio, electrical conductivity, low cost, and mechanical toughness. These diverse features enable applications ranging from energy harvesting and electronic skin to reinforced plastic materials. This review presents progress in composite materials with graphene nanoplatelets applied, among others, in the field of flexible electronics and motion and structural sensing. Particular emphasis is given to applications such as antennas, flexible electrodes for energy devices, and strain sensors. A separate discussion is included on advanced biodegradable materials reinforced with graphene nanoplatelets. A discussion of the necessary steps for the further spread of graphene nanoplatelets is provided for each revised field.

Keywords: graphene nanoplatelets; flexible electronics; wearable electronics; strain sensor; structural health monitoring; stretchable electronics; reinforced bioplastics

---

1. Graphene and Graphene Nanoplatelets

Graphene is a single freestanding monolayer of graphite [1]. It is the first 2D-material ever manufactured by mankind, having a thickness of one atom (0.34 nm), and lateral size orders of magnitudes larger [2–4]. Graphene combines diverse and unique physical properties (see Figure 1), and as a result, is an ideal building block for miniaturized next-generation devices, with applications in fields like photonics, opto-electronics, protection coatings, gas barrier films, and advanced nanocomposites [5–7].

In recent years, many studies have focused on solutions to conveniently mass-produce defect-free graphene. More than twelve different fabrication techniques were proposed [6,8–13]. Two noteworthy processes are chemical vapor deposition on copper or metals [14] and liquid phase exfoliation of graphite [15]. The first method is a bottom-up approach: it makes wide graphene films grow on top of metallic foils, starting from volatile carbon based precursors. In contrast, liquid phase exfoliation is a top-down method which singles out the graphene monolayer by sonicating graphite immersed into solvents with low surface tension or water with surfactants [15,16]. Single layer graphene flakes are then isolated only after additional ultracentrifugation steps.

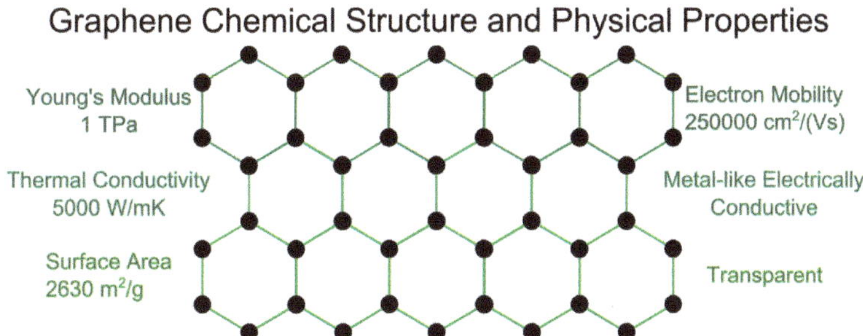

Figure 1. Graphene hexagonal honeycomb chemical structure and its remarkable physical properties. The black dots are carbon atoms.

Pure graphene is not yet mass-produced though. There is still a lack of a large scale manufacturing techniques that isolate these 2D crystals with the same outstanding performance as that required to produce the samples fabricated in research laboratories [10]. The main limitations are the low fabrication rates and high sales costs. On the other hand, graphene nanoplatelets (also known as graphite nanoplatelets, GnPs, or GPs) combine large-scale production and low costs with remarkable physical properties. This nanoflakes powder is normally obtained following the liquid phase exfoliation procedure without further centrifugation steps. Other widespread GnP manufacture methods are ball-milling [17], the exposure of acid-intercalated graphite to microwave radiation [17], shear-exfoliation, and the more recent wet-jet milling [18]. These manufacturing techniques produce a large variety of powders in terms of thickness, lateral size of the flakes, aspect ratio, and defect concentrations [18]. GnPs are composed of single and few layer graphene mixed with thicker graphite (see Figure 2); hence, structurally they are in between graphene and graphite. In literature, graphene based materials are classified according to their thickness, lateral size, and carbon to oxygen atomic ratio [19]. Considering the morphological characteristics, the graphene family can be classified as single layer graphene, few layer graphene (2–10 layers), and graphite nano- and micro-platelets. Commercially available GnPs are a mixture of single layer, few layers, and nanostructured graphite. In other words, GnPs thickness can vary from 0.34 to 100 nm within the same production batch [20,21]. Note that graphite is typically considered a 2D-like material (i.e., not bulky) when its number of layers is ≤10 [10].

GnPs exhibit exciting properties such as light weight, high aspect ratio, electrical and thermal conductivity, mechanical toughness, low cost, and planar structure. As such, they are attractive options to replace different nanostructured fillers in material science, such as other carbon allotropes (i.e., carbon black or carbon nanotubes), metallic nanoparticles, and clay [21,22]. They are appealing for nanocomposites, since they can easily and successfully be included in polymeric matrices by solvent or melt compounding [23]. GnPs are cheaper than carbon nanofibers and nanotubes, and are comparable with such tube-like nanofillers in modifying the mechanical properties of polymers [21,24]. Moreover, GnPs' electrical conductivity is orders of magnitude higher than those of graphene oxides [25].

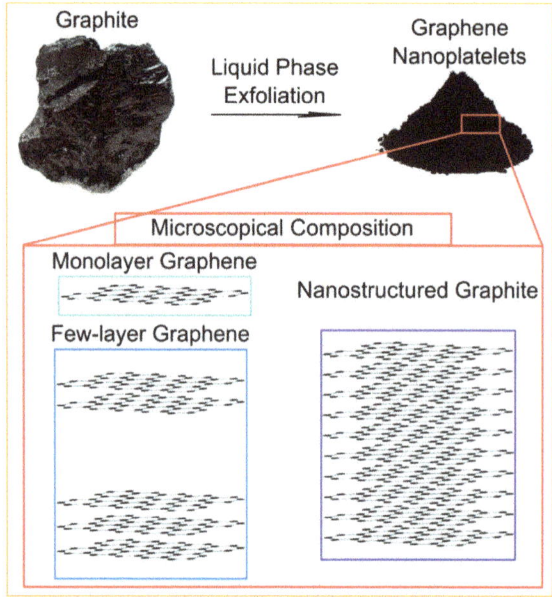

Figure 2. Schematic of the manufacture of GnPs starting from natural graphite. The typical black powder obtained after liquid phase exfoliation and solvent evaporation is constituted by a mixture of single and few layer graphene and nanostructured graphite.

Considering this, graphene nanoplatelets are already employed in several technological fields. In fact, GnPs-based materials show increased tribology [26,27], mechanical [17,28–31], biomedical [32–34], gas barrier [35,36], flame retardant [37,38], and heat conduction [39–42] properties. Furthermore, GnPs can transform plastic in an electrical conductor, converting it into a conformable material for electronics [43–45]. Finally, GnPs showed good potential for enhancing the thermal conductivity of polymer matrixes [46], making them suitable as thermal interface materials [39,47].

In this review, we will focus on GnP-based applications related in areas such as flexible and wearable electronics, motion and structural sensors, and reinforced bio-nanocomposites. In particular, we will show that GnPs unveils large-scale and unique uses (from antennas to energy harvesting) in the field of flexible electronics. We will discuss the potential uses of GnPs in smart fabrics, and the steps needed to reach a wide distribution of wearable technology. We will display many different approaches and materials employed to fabricate strain and pressure sensors, structure health monitoring systems, and stretchable devices. Finally, we will present recent advances in the field of GnP-reinforced bioplastics, and the potential of these nanoflakes to fill the performance gap between long-lasting traditional plastics and green and sustainable biopolymers.

2. Flexible Electronics Based on GnPs

Most electronic devices are based on rigid inorganic components. These conventional materials present drawbacks in light of the rise of applications that require flexibility, such as artificial electronic skin, wearable and compliant electronics, and portable energy harvesting devices [48]. The combination of the mechanical properties of polymers and conductive nanofillers is promising as a way of creating flexible and compliant conductive materials. In particular, investigation into polymers combined with silver nanoflakes showed encouraging results in flexible electronics [49]. However, nano-silver's high cost limits its large-scale production [50].

In such a context, carbon-based conductive nanofillers, and in particular, graphene nanoplatelets, gained increased attention as materials for flexible electronics due to their flexibility and low sheet resistance, i.e., that can reach the order of Ω/sq [43]. Different approaches were developed (see Table 1).

Table 1. Flexible Electronics GnPs-based. We report the manufacturing technique, electrical conductivity (EC) or sheet resistance (SR), durability tests performed and references. EC and SR are related by this formula: EC = 1/(SR $\times$ t) where t is the thickness of the material. PMMA stays for poly(methyl methacrylate), PET for polyethylene terephthalate, PTFE for polytetrafluoroethylene, PDMA for polydimethylsiloxane and PEDOT:PSS for poly(3,4-ethylenedioxythiophene) polystyrene sulfonate.

Type of Sample	Manufacturing Techniques	EC (S/m) SR (Ω/sq)	Durability Tests	Reference
Freestanding GnPs	Water dispersion and filtration	2×10^6 S/m	Not reported	[36]
GnPs-Polycarbonate Composite	Extrusion	2×10^{-6} S/m	Not reported	[51]
GnPs-Nylon 6,6 composite	Solution blending	1 S/m	Not reported	[52]
GnPs coupled with ionic liquid ions and epoxy	Solution blending and curing	10^{-3} S/m	Not reported	[53]
Polyimide substrate functionalized with GnPs	Drop casting	Not reported	Not reported	[54]
Glass, Al$_2$O$_3$ and PET substrates functionalized with PMMA-GnPs paste	Screen printing	20 kΩ/sq	Not reported	[55]
Transparent substrates coated with GnPs-PEDOT:PSS	Ink-jet printing	2×10^2 S/m	Bending (ammonia sensor)	[56]
GnPs-functionalized paper	Screen printing and rolling compression	4×10^4 S/m	Bending (antenna)	[57]
GnPs-acrylic paint emulsion on paper	Spray coating, heat-curing and polishing	5×10^2 S/m	100 abrasion and peeling	[58]
GnPs-functionalized paper	Filtration via PTFE membrane and transfer printing process	Not reported	1000 folding cycles at 180° and −180° bending angle	[59]
Deposition of GnPs on polymeric substrates, cardboard or textiles	GnPs compression with hydraulic press and lamination on different substrates	10^5 S/m	Hundreds of thousands bending cycles at bending radii of 45 and 90 mm	[60]
GnPs on PMMA with silver nanowires	GnPs brush coated on PMMA and silver nanowires sprayed on top. All the structure embedded on PET or PDMS	12 Ω/sq	100,000 bending cycles with minimum bending radius of 5 mm and stretching up to 50%	[61]
Cellulose impregnated with GnPs/Mater-bi conductive ink	Spray and Hot-pressing	10^3 S/m 10 Ω/sq	Tens of 180° folding-unfolding cycles at 0 mm bending radius.	[43]
Cellulose impregnated with GnPs/Mater-bi conductive ink	Spray and Hot-pressing.Lamination on top of a solar cell	10 Ω/sq	Solar Cell performance after bending-unbending	[62]
Cellulose impregnated with cellulose acetate and GnPs	Spray and self-impregnation	10^3 S/m 10 Ω/sq	Abrasion cycles (30 min) and tens of 180° folding-unfolding cycles at 0 mm bending radius	[63]
GnPs and nanofibrill cellulose into PLA and Polypyrrole	Solution processing	106 S/m	100 bending cycles at 180° bending angle	[64]

One method consists of fabricating freestanding GnP-based materials. Wu et al. [36] fabricated a flexible and light-weight self-standing graphene nanoplatelets paper, reaching the remarkable electrical conductivity of $\sigma \approx 2 \times 10^6$ S/m. This binder-free porous film was bent without ruptures, as shown in Figure 3. It was impregnated with both thermoset and thermoplastic polymers to increase its mechanical properties. After this impregnation procedure, the GnP paper displayed a reduced electrical conductivity ($\sigma \approx 7 \times 10^5$ S/m). Coupling with carbon fibers diminished its sheet resistance and enhanced its thermal properties. The GnPs employed by Wu et al. were prepared in their laboratory.

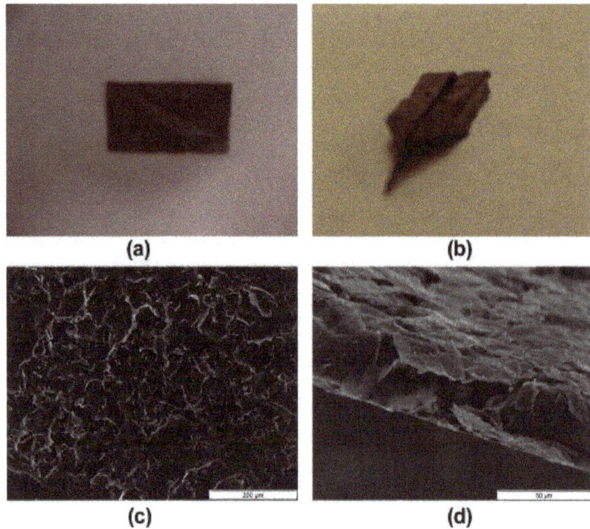

Figure 3. Self-standing GnPs paper and its flexibility. (**a**,**b**) are photographs of the paper before and after folding. (**c**,**d**) are SEM images of the morphology of the surface of the GnPs paper (plane and at the folding edge, respectively). Reprinted with permission from Carbon 50, 3, 1135–1145. Copyright 2012 Elsevier.

Although promising, the manufacture of freestanding GnPs substrates is often complicated and difficult to scale-up. Therefore, scientists explored other approaches depending on the type of polymer employed (i.e., thermoplastic or thermoset). For example, the incorporation of GnPs in thermoplastic polymer matrices led to flexible and conductive materials. Following this procedure, King and collaborators [51] extruded polycarbonate-GnPs nanocomposites with improved electrical properties. Such materials preserved ductile and plastic behavior up to 8 wt % GnPs concentration, and exhibited an electrical conductivity of approximately 2.5×10^{-6} S/m. Papadopoulou et al. [52] designed a new solvent mixture (trifluoroacetic acid and acetone) for flexible thermoplastic nylon 6.6 graphene nanoplatelets nanocomposites. They used a solvent casting method to fabricate the films. At 20 wt % nanofiller concentration, the material showed an electrical conductivity six orders of magnitude higher than that obtained by King and collaborators ($\sigma \approx 1$ S/m). They also demonstrated that, by incorporating GnPs, the pure nylon matrix improved the Young's modulus more than twice. Papadopoulou et al. employed commercially available GnPs obtained from Directa Plus (Lomazzo, Italy) (grade Ultra g+). Such GnPs were characterized in depth in our previous work [65]. Recently, Hameed and coauthors [53] proved that the use of ionic liquid induces flexibility in brittle thermoset matrices, and improves the dispersion of GnPs. Such modified thermoset polymers displayed enhanced tensile strength and Young's modulus, and were electrically conductive ($\sigma \approx 10^{-3}$ S/m).

2.1. GnPs Functionalized Substrate

Another promising approach for flexible electronics is the functionalization of bendable substrates with GnPs-based conductive ink. Tian et al. [54] fabricated temperature-dependent resistors by simple drop-casting of conductive GnPs suspensions on polyimide. Such temperature sensors were stable at high relative humidity conditions, and performed more efficiently compared to carbon nanotubes devices. Printing and spraying of conducting inks are convenient techniques to functionalize substrates, since the necessary tools are already largely diffused in the manufacturing industry [50,66]. Indeed, researchers took advantage of both methods to functionalize different flexible materials employing GnPs as conductive nanomaterials. For example, Wróblewski and Janczak [55] screen-printed flexible paste made of PMMA-GnPs, realizing electrodes on diverse substrates (glass, Al$_2$O$_3$, PET). This conductive paste, made with 1.5 wt % GnPs, had a sheet resistance in the order of 20 kΩ/sq, and transparency near 17%, enough to utilize this coating as an electrode for electroluminescent displays. Seekaew and coauthors [56] ink-jet printed conductive GnPs-PEDOT:PSS dispersion on top of a transparent substrate, manufacturing a sensor for ammonia detection. The fabrication steps and the obtained device are presented in Figure 4. The addition of only 2.33 wt % of GnPs enhanced the electrical conduction of the PEDOT:PSS conductive ink from σ ≈ 0.8 × 10^2 S/m to ≈ 1.8 × 10^2 S/m. Moreover, the sensing capability of the device was improved after GnPs addition. Indeed, GnPs enhanced the active surface area of the sample (increasing the surface roughness), and augmented the electron interaction between the sample and ammonia gas.

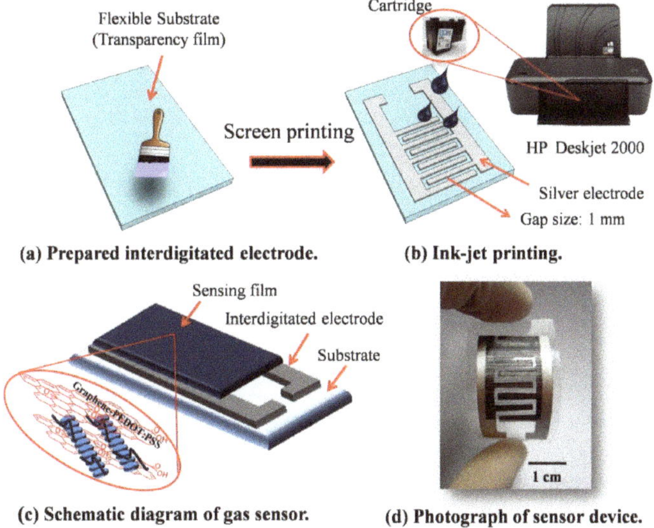

Figure 4. Schematic diagram of the manufacturing process of the ammonia sensor. (**a**,**b**) a silver interdigitated electrode was screen printed on transparent paper. The GnPs-PEDOT:PSS sensing film was deposited trough ink-jet printing; (**c**) schematic of the ammonia gas sensor. (**d**) photo of the obtained device. Reprinted with permission from Organic Electronics 15, 11, 2971–2981. Copyright 2014 Elsevier.

More recently, Huang et al. [57] used a combination of screen printing technology and rolling compression to develop GnPs-based radio frequency flexible antenna. They functionalized paper with the GnPs, obtaining electrical conductivity of 4.3 × 10^4 S/m. To verify the antenna's flexibility, they measured the reflection coefficient of bended devices, recording almost the same performance as with the un-bent antenna. To perform the described experiments, Huang et al. employed commercially available GnPs-based conductive ink (grade Grat-ink 102E from BGT Materials Ltd., Manchester, UK) which contains graphene nanoflakes, dispersants, and solvents. The described approaches result in flexible and conductive materials with remarkable applications. However, often mechanical durability and electrical features are not balanced [67]. Indeed, the lack of resistance to bend cycles and mechanical stresses limits the range of uses of such electronics materials in applications such as wearable and motile sensors technologies. Certainly, in the case of GnPs inclusion inside plastics, increasing the filler loading inside the polymer matrices can transform the latter into brittle materials and lead to complications in manufacturing [67].

Mates and collaborators [58] took one step towards the creation of GnPs-based durable materials for flexible electronics. Indeed, they realized a conducting composite coating dispersing GnPs of different sizes inside acrylic paint emulsions. Such composite films were spray casted onto Xerox printing paper, heat-cured, and polished. The adhesion of the conducting layer to the substrates was tested by Taber abrasion and peel tests, displaying remarkable resistance under such mechanical stress. The electrical conductivity reached values of approximately 5×10^2 S/m, and kept the same order of magnitude after 100 cycles of abrasion or peeling. Mates and coworkers also found that GnPs flakes with larger planar dimensions positively affect THz EMI shielding efficiency (see Figure 5). The best results obtained by Mates et al. were obtained by employing commercially available GnPs acquired from Strem Chemicals (typical thickness of 6–8 nm, lateral size of 5, 15 and 25 microns).

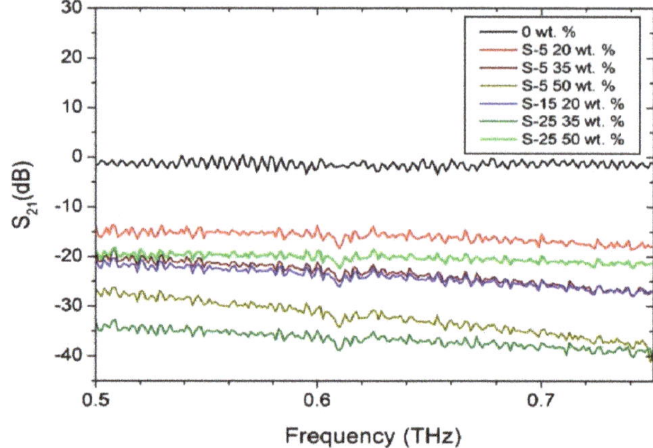

Figure 5. EMI shielding effectiveness (S_{21}) of the GnPs-acrylic paint emulsion as a function of GnPs concentration and type (S-X, where X express the average lateral size of the nanoflakes). The frequency investigated were between 0.5 and 0.75 THz. The highest level of attenuation ($\approx$36 dB) was found for the high-conductivity composites. An all-paint composite (0 wt % GnPs) was also tested as a negative reference. Reprinted with permission from Carbon 87, 163–174. Copyright 2015 Elsevier.

Another step towards reliable GnPs-based flexible electronics was demonstrated in the study of Hyun and coworkers [59]. They started by filtering a graphene dispersion using a PTFE membrane, and used a transfer printing process (a simple pen) to transfer the conductive nanoparticle onto paper. Multiple folding cycles were not sufficient to damage the material's electrical conductivity. Indeed such GnPs-paper composite maintained about 83%/94% of the initial electrical conductivity after 1000 cycles of 180°/−180° folding. Scidà and coworkers [60] designed a GnP-based antenna for near-field communication. This material exhibited significant electrical conductivity, i.e., $\sigma \approx 10^5$ S/m. The GnPs were hot-compressed, forming freestanding GnPs films that were laminated onto polymeric substrates (see Figure 6) or textiles. The performance of the devices was stable after hundreds of thousands of bending cycles at bending radii of 45 and 90 mm. The GnPs employed for this research were supplied by Avanzare (Navarrete La Rioja, España) (product AVA18, D50 = 50 μm).

Figure 6. Image of the flexible GnPs-based antenna manufactured on transparent plastic substrate. Reprinted with permission from Materials Today 21, 223–230. Copyright 2018 Elsevier.

Recently, Oh et al. [61] fabricated GnPs-based transparent electrodes for flexible optoelectronics. The nanoflakes were brush coated on PMMA and silver nanowires were sprayed uniformly on top. The entire structure was embedded onto PET or PDMS. With this technique, a sheet resistance of 12 Ω/sq with transmittance of 87.4% was reached. After 10^5 bending cycles, the resistance increased by only the 4%. Such a GnP-based electrode was doped with p-type $AuCl_3$ and Cl_2, and used as the anode in organic light emitting diodes, substituting and performing better under bending and stretching than standard indium thin oxide.

2.2. Environmentally-Friendly Graphene-Based Materials and Devices

Another valuable and important parameter for the electronics of the future will be their sustainability (i.e., the biodegradability of the components and/or the green approaches employed to produce the materials) [68,69]. Indeed, electronic goods production and waste management have become a major issue for environmental pollution [68,69]. A novel method was proposed by our group [43] to fabricate isotropically electrically conductive biodegradable biocomposites based on cellulose and GnPs. It consisted of hot-press impregnation of porous cellulose networks after spray coating the flexible fibrous cellulose substrates with conductive GnPs-based inks. Since such ink was made employing a biodegradable thermoplastic polymer (Mater-Bi®), hot pressing at a temperature higher than the melting of the plastic led to the polymer-GnPs incorporation inside the fibrous network. The resultant green materials exhibited remarkable electrical conduction ($\sigma \approx 10^3$ S/m) and a significant folding stability after severe weight-assisted 180° folding-unfolding cycles at 0 mm bending radius. Such conductive materials were used to fabricate simple circuitry [43], and as a top electrode for organic photovoltaics solar cells [62]. Another green approach developed by our group [63] to obtain reliable bio-based material for foldable electronics was to take advantage of the liquid absorbing properties of pure cellulose. A green conductive ink realized employing methanol and acetic anhydride as solvents, and cellulose acetate and GnPs as solid content, was spray coated onto pure cellulose. The ink thoroughly wet and impregnated the cellulose substrate after deposition, eliminating the need for hot-pressing. This cellulosic-GnPs bionanocomposite exhibited good folding stability and abrasion resistance. Proposed applications were sustainable THz electromagnetic shielding materials and electromyography signal detection (see Figure 7). The GnPs employed by our group for these studies were provided by Directa Plus (grade Ultra g+). For details on the lateral size and thickness of such nanoflakes, see this report [65].

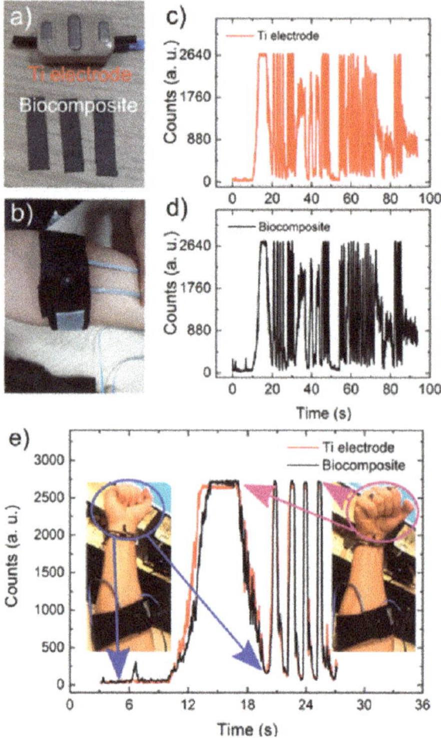

Figure 7. GnPs-cellulose nanocomposite used as electrode for surface electromyography. (**a**) Photograph displaying a standard electromyography titanium electrode (Ti electrode) and the cellulosic biocomposite; (**b**) Photograph showing both electrodes strapped to the arm side by side; (**c**) Signal acquired from the titanium conductor; (**d**) Signal recorded from the biocomposite; (**e**) Superimposed signals from both conductors. The signal rests at zero when the wrist is not flexed (inset on the left). The signal saturates when the wrist if flexed (inset on the right). Reprinted with permission from Advanced Electronic Materials, 2, 11, 1600245. Copyright 2016 Wiley.

Another green nanocomposite for flexible electronics was lately proposed by Liu and collaborators [64]. They added GnPs and cellulose nanofibril into polylactic acid and conductive polypyrrole composite via a green, cost effective method. The addition of GnPs enhanced the electrical conductivity of the biocomposites from 12 to 106 S/m at 10 wt % nanoflakes concentrations. The nanocomposites exhibited also remarkable flexible stability, with only 7.5% deviation after 100 cycles. The GnPs and nanofibril addition also enhanced the mechanical properties of the biocomposite. The nanocomposite was employed as the electrode for flexible supercapacitors.

2.3. Flexible Electronics Outlook

In brief, GnPs have a high potential for flexible electronic devices. EMI shielding, antennas, supercapacitors, and bendable electrodes for solar cells are the most promising applications. The employment of large scalable production processes, together with the industrial availability of GnPs and the high resistance to mechanical stresses (e.g., bending cycles and abrasion), are all encouraging for the large-scale expansion of GnP-based flexible electronics. The large number of work dealing with the functionalization of cellulose and/or the employment of biopolymers and green methods is also promising for the future of sustainable electronics [68,69].

3. Wearable Electronics Based on Graphene Nanoplatelets

Smart textiles are fibrous materials with numerous functionalities and applications compared to common fabrics [70]. It is predicted that the wearable device market will reach US$ 20.6 billion in the current year (2018) [71]. Electrical conductivity is the main promising feature of these garments, because it is highly sought after in fields such as flexible, wearable, and deformable electronics, and for the emerging Internet of Things [71,72]. Miniaturization propelled by nanotechnology allows us to fabricate electronic components, even working on a single fiber [72]. However, direct functionalization of textiles is more often targeted by the incorporation of conductive nanoparticles inside/on fibrous networks [73–76]. Using such an approach, remarkable results were achieved in applications related to generators [77], supercapacitors [78], and electrochemical sensors [79]. Nevertheless, further research is needed to create wearable conductive materials with stable electronic performance under mechanical stress and ambient conditions (e.g., sunlight, air etc.). In particular, wearable electronics require the creation of a new class of materials which are flexible, foldable, and washable, and which, at the same time, maintain a satisfactory level of electrical conductivity [48]. Polymeric nanocomposite materials, due to the intrinsic mechanical properties of polymers and ease of manufacturing, and to the large spectra of properties accessible with different nanoparticles, are suitable for conductive wearable technologies [80].

In this regard, promising approaches are: (1) the functionalization of fabric with Graphene Oxide (GO) [76,81–83], (2) the employment of graphene-based materials to produce conductive fibers [84–86], and (3) the transfer of chemical vapor deposited graphene films onto textiles [87,88]. All these approaches present enhancements compared to metal functionalization [89]. However, some limitations can be identified [89]:

(1) Graphene oxide needs reduction steps, and the obtained sheet resistance is often high (i.e., in the order of thousands of Ω/sq).
(2) Graphene freestanding fibers have remarkable electrical properties, but difficult adaptability to the current garment industry.
(3) Chemical vapor deposition of graphene is expensive, and the transferring procedure of the film is complicated.

Recently, another promising procedure was reported to impart electrical conductivity to industrially produced fabrics through a functionalization based on GnPs. This method has the advantage of being adaptable to several commercially-available materials like cotton and polyesters, and since GnPs are already produced in amounts suitable for the textiles market (hundreds of tons), it is scalable.

For example Woltornist et al. [89] prepared a conductive GnP-infused poly(ethyleneterephthalate) fabric by dip coating and tip sonication in heptane and water. They reached a few kΩ/sq sheet resistances at 15 wt % loadings. Sloma and coauthors [90] fabricated electroluminescent structures on textiles (paper and cotton), employing GnPs as transparent electrodes. They obtained a transmission of 70% of the incoming light, and sheet resistances in the order of 10 kΩ/sq. Tian and collaborators [91] produced a conductive fabric by layer-by-layer deposition of GnPs doped PEDOT:PSS-chitosan on cotton. Such fabrics featured ohmic I-V curves and could achieve an electrical conductivity of $\sigma \approx 0.4$ S/m (see Figure 8a). Furthermore it was resistant to washing cycles, and simple circuitry could be realized, powering up LEDs, as shown in Figure 8b. The fabric also showed remarkable ultraviolet protective ability, i.e., approximately 300-fold higher than the control fabric.

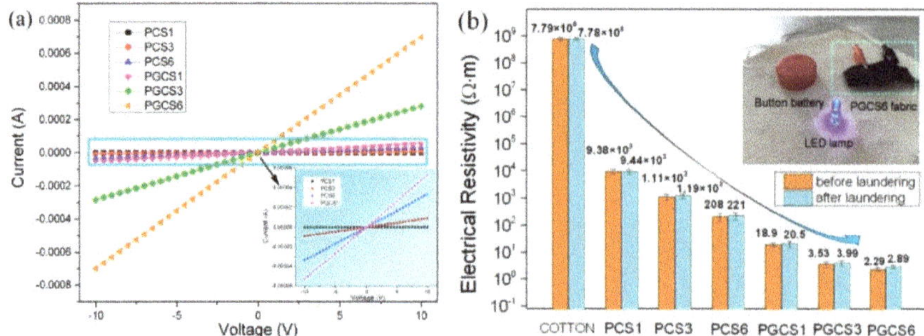

Figure 8. Electrical characteristics of the conductive cotton fabric produced by layer by layer deposition of GnPs doped PEDOT:PSS-chitosan on the textile. (**a**) I–V curves for different specimen labeled as PCSX (PEDOT:PSS and chitosan) and PGCSX (PEDOT:PSS/GnPs and chitosan) were X is the number of layer deposited. For example PCS1 is the sample obtained depositing one thin film of PEDOT:PSS and chitosan; (**b**) Electrical resistivity of control fabric, PCSX and PGCSX fabrics (before and after water laundering 10 times). Reprinted with permission from Carbon 96, 1166–1174. Copyright 2016 Elsevier.

Printing and spraying were also employed in the context of wearable electronics. For example, Skrzetuska et al. [92] screen printed a GnP-carbon nanotubes conductive paste onto cotton. The formulation of the paste was water based, and they were able to reach sheet resistances in the order of a few kΩ/sq. To bind the textile and the conductive nanofillers, they added a cross-linking agent (aliphatic urethane acrylate) to the preparation. Recently, our group [93] fabricated wearable conductive cotton fabric through simple spray procedures. A conductive ink was realized by mixing thermoplastic polyurethane (TPU) and GnPs. Such a dispersion was deposited onto cotton and impregnated through a hot-pressing procedure. The resulting material exhibited sheet resistances ~10 Ω/sq, as shown in Figure 9a. Furthermore, the conductive fabric displayed significant resilience against multiple weight-pressed folding cycles, while folding-induced micro-cracks could be easily healed by repeating the hot-pressing procedure, restoring the initial value of sheet resistance (Figure 9b). The nanocomposite conductivity was unaffected by high humidity conditions and solar irradiation, and was slightly modified by laundry cycles. In our research we used a commercially available GnPs (grade Ultra g+ GnPs, Directa Plus S.P.A.).

Wearable Electronics Outlook

As shown, GnPs have good potential in smart fabric applications. Nevertheless, so far, the highest limitation is the difficulty to bind GnPs to textiles and ensure washing stability and durability. In this prospect, GO is more diffused because, with its oxygen groups, it is simpler to be chemically linked to cotton [76]. However, GnPs are more suitable than GO for applications which need high electrical conductivity. Therefore, smart solutions to effectively bind the nanoplatelets with fabrics will enlarge their potential in wearable and textile electronics, and boost the spread of GnP-based smart garments. The junction of nanoflakes with elastomeric polymer seems to be a suitable approach [93].

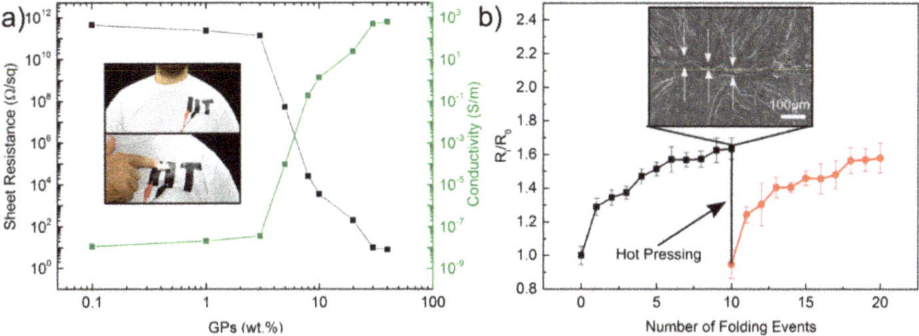

Figure 9. Electrical characteristics of the thermoplastic polyurethane GnPs nanocomposites spray coated on cotton. (**a**) Electrical conductivity and sheet resistance measurements as a function of GnPs concentration with respect to thermoplastic polyurethane. In the inset, a conductive path is spray coated on a t-shirt and used as conductor to light-up LEDs. (**b**) Black line: modifications in the sheet resistance due to weight-pressed 0 mm bending radius folding-unfolding events. The inset displays crack formation after the 10th cycle. Red line: performance after healing the crack by hot-pressing. Reprinted with permission from ACS Appl. Mater. Interfaces 9, 16, 13825–13830. Copyright 2017 American Chemical Society.

4. Graphene Nanoplatelets for Strain Sensors and Stretchable Electronics

The field of wearable and stretchable electronics pushes towards unusual material arrangements to design circuitries on curvilinear and deformable structures/organisms, and fabricate soft electrical devices and sensors [71,94–97]. The junction of a pliable and elastomeric material with a conductive nanomaterial is a favorable approach to build compliant nanocomposites for electronics [98–100]. Sensor technologies exploit conductivity and capacitance fluctuations recorded when such nanocomposites are deformed (stretched, wrapped, or compressed). These variations are used as a feedback mechanism for devices like strain, pressure, and tactile sensors [101–103]. Typically, with an external stimuli (often mechanical), conductive nanofillers inside the matrix are separated apart or connected further, changing the material's electrical features.

Following such approaches, flexible and stretchable GnP-based sensors were investigated and designed in the past years (see Table 2) [48].

Table 2. Smart Sensors GnPs-based. Gauge factor is defined as the ratio between electrical resistance change and mechanical strain.

Type of Sample	Manufacturing Techniques	Type of Sensor	Characteristics	Reference
GnPs on PET	Spray coating	Piezoresistive Strain Sensor	Gauge factor 150	[104]
GnPs or CnFs on nitrile rubber	Spray coating	Stretchable Tactile Sensor	Sensitivity 0.03 N	[105]
GnPs-PEDOT:PSS on cotton	Spray Coating	Piezoresistive Strain Sensor	Gauge factor 5	[106]
GnPs inclusion in epoxy resins	Solution processing	Piezoresistive Strain Sensor	Gauge Factor 750	[107,108]
GnPs inside PDMS	Solution processing and molding	Piezoresistive Strain Sensor	Gauge factor 230	[109]
GnPs inside Silly Putty	Solvent mixing	Piezoresistive Strain Sensor	Gauge factor > 500	[110]
TPU-GnPs nanocomposite	Solvent Mixing	Piezoresistive Strain Sensor	Adjustable electrical properties	[111]
GnPs-CnTs inside PDMS	Screen Printing	Piezoresistive Strain Sensor	Gauge factor 100	[112]
GnPs-CnT dispersed in PMMA/PVDF	Screen printing on plastic	Pressure Sensor		[113]
Glass fiber coated with GnPs	Dip coating	Piezoresistive Strain Sensor	Gauge Factor 16,000	[114,115]
Textiles functionalized with GnPs-poly(vinyl alcohol)	Dip Coating layer by layer assembly	Piezoresistive Strain Sensor	Gauge factor 1800	[116]
GnPs on medical tape and embedded in PDMS	Press and molding	Piezoresistive Strain Sensor	Gauge factor 110	[117]
CnTs grow on GnPs inside PDMS	Mechanical Mixing	Piezoresistive Strain Sensor	Gauge factor 1000	[118]
GnPs-PDMS nanocomposites	Layer by layer spin coating	Capacitive Strain Sensor	Linear capacity variation	[119]
GnPs-PDMS foam	Direct template	Pressure Sensor	Sensitivity 0.23 kPa^{-1}	[120]

Strain sensing is crucial for the advances in smart robots, human/structure health monitoring, and human-machine interactions [108,117,121,122]. Indeed, the strain gauge market surpassed 4.5 billion $ in 2013, and is growing constantly [104]. The most important parameter for strain sensing is the gauge factor. This parameter is defined as $(\Delta R/R)/(\Delta L/L)$, where $(\Delta R/R)$ is the relative change in electrical resistance (R) obtained under material elongation, and $(\Delta L/L)$ is the applied strain. Graphene-based materials have shown gauge factors among the highest ever reported [103]. Two-dimensional nanoflakes usually show a piezoresisitivity one order of magnitude higher than that of nanowires, since their electrical percolation network is largely susceptible to geometrical changes and discontinuities [123]. Furthermore, carbon-based fillers can expand the utilization of strain sensors, enhancing their elongation range from a few to several hundred percent stretches [122,124].

Different approaches were employed to realize GnPs-based strain sensors. The most commonly-used feedback mechanism is based on stretch-induced electrical resistance changes (piezoresistivity). Hempel et al. [104] designed strain gauges by simply spraying pristine GnPs thin film on plastic PET substrates. Changing the spray parameters, they were able to control the coating morphology and decrease the film resistance by varying the amount of deposited dispersion. The gauge factor was stable over 4000 strain cycles, and exceeded 150. The electrical percolation behavior of the system under stretch was in good agreement with a model simulating a link between randomly-oriented conductive disks.

4.1. PDMS and Graphene Nanoplatelets for Strain Sensing

A widespread material for strain sensors is polydimethylsiloxane (PDMS), due to its flexibility, stretchability and ease of manufacturing [125–130]. PDMS-based strain sensors were also investigated coupled with GnPs. For example, Wang and collaborators [109] developed GnPs-PDMS composites by simple sonication and molding processes. This device reached gauge factors of approximately 230 at a GnPs concentration of around 8 vol %, and within a strain of 2%. Shi and collaborators [117] produced an electrically conductive and stretchable film by mechanically pressing GnPs onto a medical tape and embedding the structure inside PDMS. The manufacturing was low-cost and fast (in the order of one minute), and therefore, large-scalable. The obtained sensor was reliable over 1000 stretch-release cycles, exhibited a time of response of less than 50 ms, and a gauge factor of approximately 110. Moreover their sensor showed sensitivity to low mechanical strains. Indeed, it could detect minuscule movements from a cricket, and air vibration caused by mobile phone speakers, as shown in Figure 10.

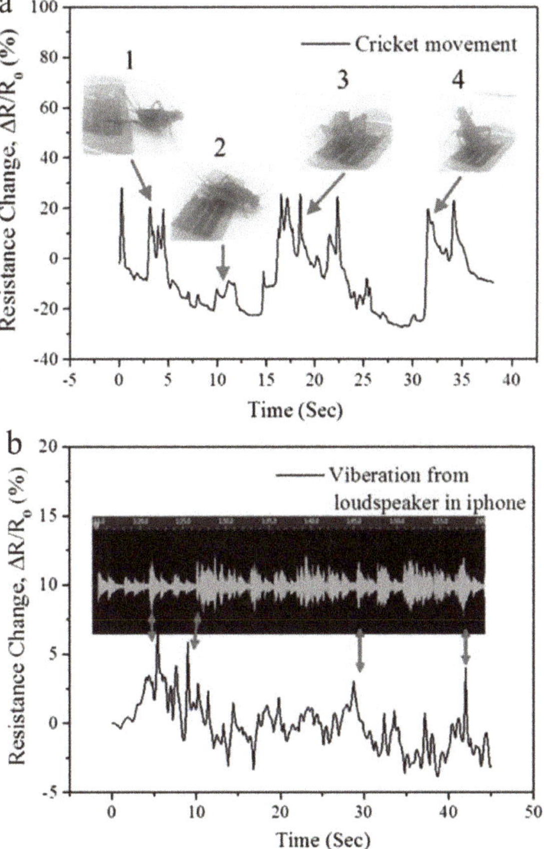

Figure 10. Performance of the deformation sensor obtained pressing GnPs on a medical tape and embedding such structure inside PDMS. The sensor was employed as tiny movements and sound signal detector. (**a**) Signal generated by the footstep of a field cricket moving on the sensor and (**b**) recognition of music signal from an iPhone speaker, with inset being audio signal. Reprinted with permission from Advanced Functional Materials. Copyright 2016 Wiley.

Another PDMS and GnPs composite for strain sensing was investigated by Lee and collaborators [131]. They used layer by layer assembly to create controlled geometries of GnPs on the elastomeric substrates. Such devices can monitor subtle human movements.

The synergic properties of GnPs and carbon nanotubes are convenient to design PDMS-based strain sensors. For example, Lee et al. [112] screen printed a biocompatible gauge device employing multilayer graphene nanoflakes and multiwall carbon nanotubes (MWCNT). The gauge factor changed from 22 (for 12.5 wt % GnPs inclusion) to 100 for concentrations near the percolation threshold (i.e., 1.5 wt % GnPs and 3.5 wt % MWCNT blend). The authors attribute such behavior to the large contribution of electron tunneling at nanofiller loads close to the percolation threshold; with stretch, the distance between nanofiller units augment, however, still permitting tunneling conductivity. For higher filler loads, the mechanism of conductivity is mostly determined by direct contacts that are slightly affected by strain compared to tunneling electrons. More recently, Zhao and collaborators [118] used catalyst chemical vapor deposition approaches to grow nanotube forests on both sides of GnPs nanoplatelets. The nanoflakes-nanowires filler was mechanical mixed with PDMS, achieving percolation threshold at 0.64 vol %. At 0.75 vol % concentration, they reached a gauge factor in the order of 1000. Such sensors are able to sense and even distinguish between tiny finger motions, as shown in Figure 11. Furthermore, the nanocomposites were also used as compression sensors, reaching pressure sensitivity of 0.6 kPa^{-1}.

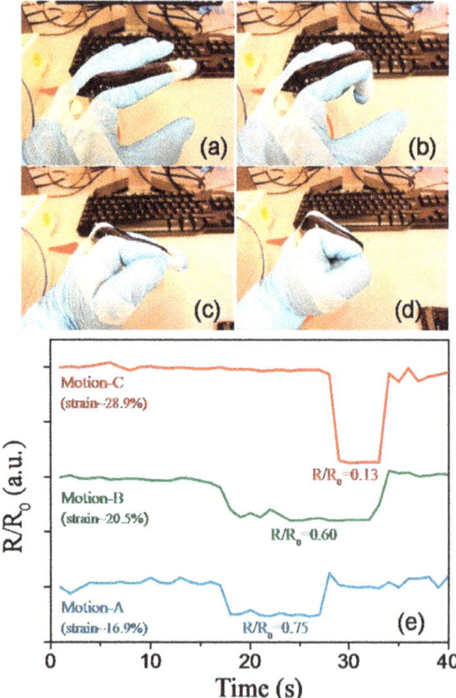

Figure 11. Nanoflakes-nanotubes PDMS composite film glued on rubber gloves and serving as finger motion detector. (**a**) Photograph of the relaxed state of the finger; (**b–d**) are photographs of second and third joint and clenching motions respectively; (**e**) Resistance changes for each independent movement. Reprinted with permission from ACS Appl. Mater. Interfaces 7, 9652–9659. Copyright 2015 American Chemical Society.

4.2. Other Approaches for Graphene Nanoplatelets Based Strain Sensing

Apart from PDMS, other rubbery materials are also candidates for possible implementation for strain sensing. For example, Boland and coauthors [110] used a highly viscoelastic, lightly cross-linked silicone polymer (trademark name Silly-Putty®) mixed with GnPs to design a material that can monitor impact, pressure, and deformation. Its electrical resistance augmented linearly when stretched below 1%, and surprisingly decreased, even below its initial value for elongations between 1% and 10%. Such a trend is unique considering the monotonic increases of resistance usually observed with stretch. Silly-Putty-GnPs nanocomposites exhibit gauge factors of around 500 at 6.8 vol % of nanofillers. Such device can measure blood pressure, pulse rate, and even the footstep impact associated with spiders. The GnPs (Lateral size $\approx$ 500 nm, thickness $\approx$ 20 layers) employed for this work were prepared by ultrasonic tip-sonication of graphite (Branwell, Graphite Grade RFL 99.5) in NMP. These GnPs were then dispersed in chloroform after NMP drying. Recently, our group [111] showed that electrical conductivity of TPU-GnPs nanocomposites can be tuned and improved by repeated stretching cycles, without exceeding 20% of the maximum strain. A decrease of up to 60% in electrical resistance was measured after 1000 stretch-release cycles. We discovered that the described changes were caused by stretch-induced redistribution of the GnPs within the polymer matrix. Such TPU-GnPs nanocomposites can be used for strain sensing applications after stretch-induced electrical feature optimization. The rearrangement of the disposition of nanoflakes inside TPU was noted also by Liu and collaborators with single layer graphene [132]. Wearable and stretchable textiles are ubiquitous. Another common strategy for strain sensing is the functionalization of such fabrics with conductive materials. For example, Zahid and collaborators [106] obtained a conductive ($\approx$200 S/m) textile functionalizing cotton through a simple spray of PEDOT:PSS-GnPs dispersions. The material exhibited strain sensing capabilities, with a gauge factor of approximately 5 at strain of 5% and 10%. It could resist repeated washing and bending events, thereby ensuring possible commercialization.

The combination of high sensitivity to tiny deformation and broad sensing range is often unusual in the field of strain sensors [133]. In such a context, promising results were obtained by Park et al. [116], who reported on the fabrication of stretchable yarns realized through simple layer-by-layer assembly. In particular, different yarns (rubbery, nylon-rubbery and wool) were immersed first in a poly(vinyl alcohol) solution, and, when dry, in a GnPs dispersion. All these fabrics maintained a remarkable stretchability (see Figure 12). When needed, a PDMS coating was used to prevent GnP detachment during stretch. Depending on the fiber used, different performances were achieved. Indeed, the highly sensitive rubbery sensors exhibited a gauge factor of $\approx$1800 and a maximum stretchability of around 100%, the nylon-rubbery device led to a gauge factor of 1.4 and a maximum elongation of 150%, and the wool-based sensor displayed an atypical negative gauge factor of -0.1 with a maximum stretch of the 50% of the initial length. These performances are suitable for monitoring a wide range of body movements depending on the fiber employed, from tiny breathing to finger-bending.

Moriche et al. [114] investigated the use of fabric-like material coupled with GnPs for strain sensing applications. They dip coated glass fiber with conductive nanoflakes, observing that this coating was more effective when GnPs were NH_2 functionalized. Also, electrical conductivity and strain sensing were strongly enhanced when GnPs were doped with nitrogen. Such GnP textiles exhibited an exponential increase in resistance with stretch, reaching gauge factors in the order of 16,000. Additionally, functionalized fibers attached on nitrile gloves were able to monitor single finger movements. Strain sensing can be also performed by monitoring the electrical capacity variation when a material is mechanically deformed. Filippidou et al. [119] realized a deformable PDMS-GnPs plain capacitor. They used a GnP-PDMS composite as the soft electrode and pure PDMS as the elastic dielectric. The capacitor was manufactured simply by a layer-by-layer spin coating technique. The sensor was tested for small strain measurements in the order of 0.2%, showing a linear variation of capacity with stretch.

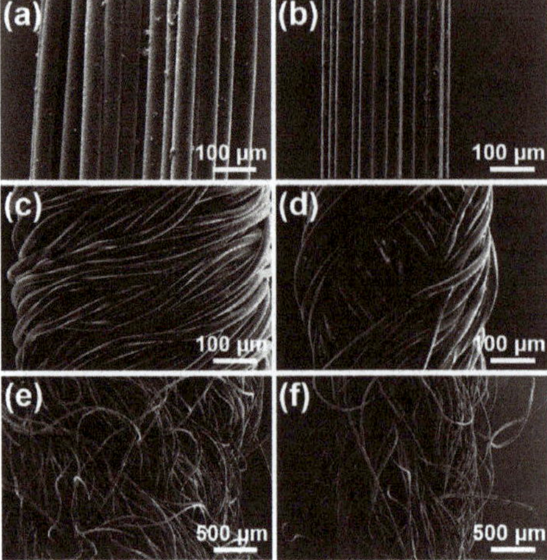

Figure 12. Stretchable and conductive yarns realized through simple layer by layer assembly on different elastomeric textiles. (**a,c,e**) Scanning electron microscope image of the GnPs strain sensors realized with rubber, nylon-rubber and wool fibers, respectively; (**b,d,f**) the same textile stretched (100%, 100% and 50% respectively). Reprinted with permission from ACS Appl. Mater. Interfaces 7, 11, 6317–6324. Copyright 2015 American Chemical Society.

4.3. Graphene Nanoplatelets Based Structural Health Monitoring

Structural health monitoring (SHM) is a specific strain-sensing application which needs extremely high sensitivity to minute structural changes [134]. It was thoroughly investigated to enhance the safety of buildings and monitor the stability of structures in real-time, enhancing public security. Due to the high gauge factor, GnP-based nanocomposites are adequate for this purpose. Indeed, recently, researchers have focused on such devices. For example, Moriche and collaborators [107] added GnPs to an epoxy resin used for aeronautic applications. Nanocomposites with nanofiller contents of around the electrical percolation threshold displayed high gauge factors ($\approx$750) at strains inferior to 1%. They discovered that the dominant strain sensing mechanism was based on changes on the distance between nanoflakes. The resultant exponential electrical change was explained with a diminished tunnel effect. Increasing the amount of GnPs, the direct contact mechanism dominated, and the electrical response with stretch became linear. They were able to discriminate between structural changes due to tensile or flexural stress. They also discovered that such modified epoxy resin-GnPs present no hysteresis during 50 cycles of loading-unloading in flexural test conditions, demonstrating the reversibility of the SHM mechanism in the plastic deformation regime [108]. Recently, the same group [115] performed a study on SHM materials obtained by combining NH_2 functionalized GnPs, epoxy resins, and glass fiber. The experiments performed in this section were performed with commercially available GnPs provided by XGScience (nominal thickness 6–8 nm and lateral size 25 microns).

4.4. Pressure Sensors

In addition to strain sensors and SHM, Rinaldi et al. [120], advanced flexible and compressible pressure sensors functionalizing PDMS foams with GnP dispersions. The manufacture of PDMS foam was achieved following the direct template technique, which implicates the replication of the inverse assembly of a preformed leachable prototype (see Figure 13 for details). In particular, they discovered that such device exhibited a linear regime for pressures <10 kPa, while at higher pressures, the electrical conductivity increase with higher steep (maximum 70 kPa, 800% resistance change). The sensitivity was of 0.23 kPa^{-1}.

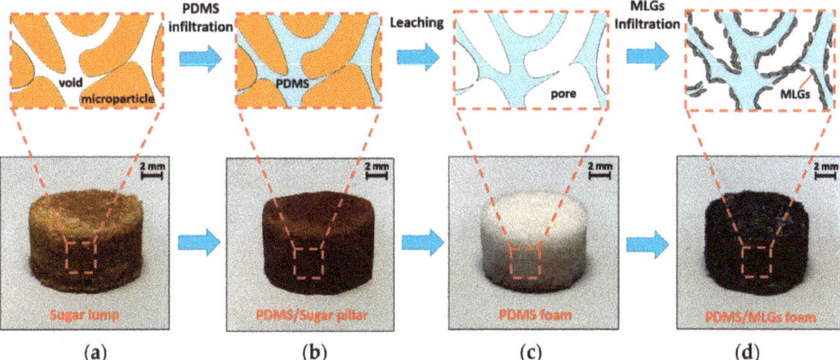

Figure 13. Manufacture steps for the fabrication of PDMS-GnPs foams. (**a**) Sugar template; (**b**) PDMS infiltrated sugar template; (**c**) Sample after the sugar removal; (**d**) PDMS/Multilayer graphene (MLGS) foam resulting after nanoflakes infiltration. Reprinted with permission from Sensor 16, 12, 2148. Copyright 2016 MDPI.

Janczak and collaborators [113] screen printed GnPs and carbon nanotubes dispersed in PMMA or PVDF on flexible substrates. Such conductive composites were employed as the active layer in pressure sensors. The sensitivity of the sensor was proportional to the sheet resistance of the material.

4.5. Capacitive Sensors for Tactile Sensing

There are applications in stretchable electronics which require materials with stable electrical features under constant elongation conditions, such as electronic skin and elongating touch sensors. One of the methodologies employed to build artificial electronic skins is deformable capacitors that can identify touch-induced pressure, shear, and torsion [135]. Ensuring the functionality of such capacitive devices under mechanical stress (bent and even more stretched exceeding 100%) is challenging [136].

Our group [105] recently reported the fabrication of a durable stretchable haptic capacitive sensor using nitrile rubber as template. A conductive elastomeric polymer dispersion containing GnPs or carbon nanofibers (CnFs) was spray coated onto both sides of a nitrile rubber piece, obtaining a parallel-plate capacitive touch sensor. The conductive spray, either GnP- or CnF-based, reached satisfying sheet resistance levels, i.e., $\approx$10 Ω/sq. The GnP-based conducting electrodes formed cracks before 60% elongation, while the conducting electrodes based on CnFs sustained their conductivity at up to 100% strain level. However, both electrodes were adaptable and trustworthy, considering the motility and elongation level of human junctures ($\approx$20–40% strain). Remarkably, structural deterioration due to cyclic stretch-release events could be healed as a consequence of a straightforward heat gun annealing process. We also demonstrated the haptic sensing characteristics of an elongating capacitive device by wrapping it around the fingertip of a robotic hand. Tactile forces could be detected without difficulty by the device over curvilinear surfaces or under elongation (see Figure 14). The experiments performed in this section were performed with Ultra g+ commercially available GnPs obtained by Directa Plus S.P.A.

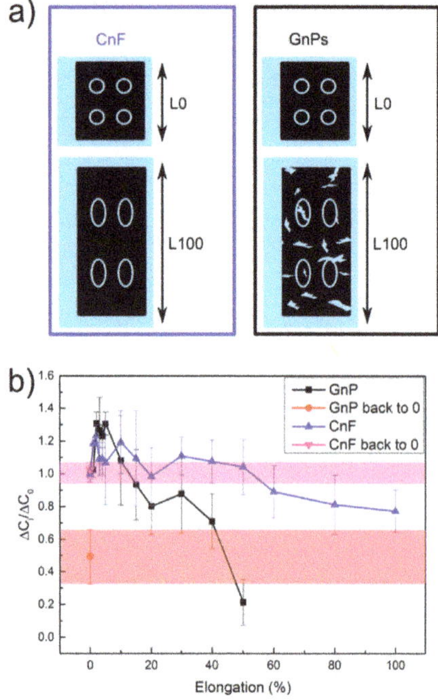

Figure 14. Haptic sensor functionality under elongation of elastomeric conductive inks sprayed on both sides of nitrile glove pieces. (**a**) Graphic representation of the CnF- and GnP-based tactile sensor at 0% elongation (L0) and at 100% stretch (L100); (**b**) device performances at consecutive elongation steps. ΔC_i and ΔC_0 represent the capacity deviation with touch under elongation and at 0% stretch, respectively. Reprinted with permission from Advanced Science 5, 2, 1700587. Copyright 2017 Wiley.

4.6. Smart Sensing Outlook

In summary, GnP-based smart sensors have been extensively investigated to date and are already considered mature for large-scale production. In particular, strain sensing is a more promising application, considering the high gauge factor obtained using GnPs. Such nanoflakes are an excellent candidate for applications which demand high sensitivity to tiny deformations (like structural and health parameter monitoring), broad sensing range, or a combination of the two. To date, PDMS-GnPs strain sensors are the most studied. Another promising approach for stretch sensing is functionalizing textiles with conductive materials to obtain fibrous strain sensors. So far, piezoresistive devices were investigated much more, but some examples using capacitive feedback mechanism are present in the literature. GnP-based stretchable and deformable sensors were also employed for pressure sensing and robotic tactile sensing, obtaining remarkable results.

5. Advanced Reinforced Graphene Nanoplatelet-Based Bio-Nanocomposites

Graphene-based reinforced nanocomposites showed a prominent role in the field of advanced materials [137–141]. The outstanding mechanical properties of single layer graphene, GnPs and GO, led to a new generation of improved plastic-based structural materials [17,142,143]. For example, these 2D carbon-based fillers are attractive for the realization of next generation sporting goods [144], concrete [145], anti-corrosion coatings [146], automotive lightweight components [147], structural elements in aerospace [148], and wind turbine designs [149]. In particular, GnP powder is already produced at large scales and at a low price compared with single-layer graphene, and as such, is more appealing for the composite market [150]. Furthermore, GnPs were already selected as nanofillers for toughening polymeric materials, since they are strong, and present lower defect concentrations compared to GOs [151]. GnPs integrate the excellent mechanical characteristics of carbon nanotubes with the multi-flakes structure of clays, which can impart superior structural property improvements [150]. Indeed, they were shown to perform significantly better than these nanofillers in enhancing the mechanical properties of nanocomposites such as tensile strength, elastic modulus, fracture toughness, fracture energy, and resistance to fatigue and crack growth [65,152], due to their higher compatibility with polymers matrix [153].

To obtain an effective reinforcement, a crucial requirement is the homogeneous dispersion of the flake-like filler inside the polymer matrix [17,24,143,154–156], as schematically shown in Figure 15a. However, such conditions are necessary but not sufficient. Indeed, also flake dimensions (lateral size and thickness), alignment, and chemical interaction with the matrix (see Figure 15b) need to be taken into account [24,154,156–158]. Furthermore, a recent paper by Papageorgiou et al. [159] investigates in depth the mutual interaction between graphene-based materials and the hosting matrix. Through the examination of hundreds of papers on graphene composites, they demonstrated that the fillers modulus is dependent on the polymer matrix. In particular, the GnPs modulus is larger when the filler concentrations are reasonably low and the matrix is more rigid. This implies that the common hypothesis that the nanoflake modulus is independent of the matrix is not correct.

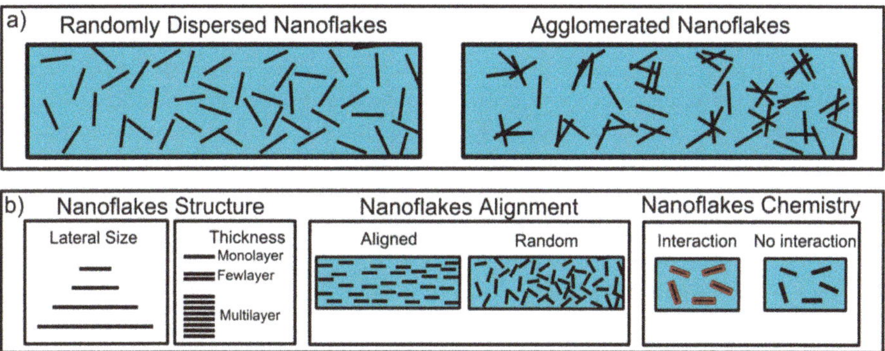

Figure 15. Important parameters for polymer reinforcement by nanoflake inclusions. (**a**) Schematic of homogeneously dispersed and agglomerated filler inside a polymer matrix. Agglomerations enhance crack propagation; (**b**) Schematic of additional significant parameter for effective reinforcement of plastics with graphene based nanofillers: nanoflake structure, alignment, and chemical interaction with the matrix.

Many works have been published so far dealing with the incorporation of graphene, graphene oxide or GnPs inside traditional synthetic and long-lasting plastics. Several reviews are already available on these topics [17,20,21,23,28,137–141,144,153]. In contrast, reviews centered on the reinforcement of bioplastics with graphene-based materials are still rare [160].

Biopolymers are biodegradable; thus, they contribute only slightly to environmental pollution [161,162]. They are constantly gaining interest, considering that they could gradually substitute oil-based, polluting, synthetic polymers [163–165]. Nevertheless, a broad spread-out of these green plastics is often limited, taking into account their poor mechanical properties [162–164]. Therefore, the incorporation of nano-sized reinforcements in bio-polymeric matrixes is emerging as a strategy to improve the performance deficiency of the bioplastics, and obtain characteristics comparable with the traditional long-lasting ones [163,164,166,167]. As such, the inclusion inside biodegradable matrixes of different nanofillers, such as nanostructured metals, multilayered silicates, silica nanoparticles, and carbon nanomaterial were extensively investigated [163,164,166–168]. Single-layer graphene and GO were used coupled with biopolymers, and two reviews were written on this topic [167,169]. In contrast, to the best of our knowledge, there is not a systematic review dedicated to GnP-reinforced bioplastics, although their potential in this field is steadily growing (see Table 3) [167].

Table 3. Biocomposite GnPs reinforced. The Young modulus is named E_g, the tensile strenght T_s and the elongation at break S_g. These properties are expressed in percent increment. "=" means unchanged.

Matrix	Manufacturing Techniques	E_g	T_s	S_g	Comment	Reference
BioFlex®	Melt Blending	40%	N/A	N/A	at 5wt % filler	[166]
PLA	Melt Blending	12	20	16	0.25 wt % filler. GnPs inside PLA did not affect human fibroblasts morphology and metabolic activity	[170]
PLA or BioFlex®	Melt Blending	40	=	=	5 wt % filler. GnP affected release of ciprofloxacin without preventing the antimicrobial activity.	[162,171]
PLA plasticized with palm oil Pla/Poly(ethylene Glycol)/palm oil	Melt Blending	N/A	27	60	0.3 wt %. Increased antibacterial properties	[172] [173] [174]
PLA	Melt Mixing	200	N/A	N/A	3 wt %	[175,176]
PLA	Melt compounding	Large 24 Small 10	N/A	N/A	5 wt % Study on the effect of GnPs size on PLA mechanical properties	[177]
PLA	Solution Processing Melt Blending	GO 115% GnPs 156%	N/A	N/A	0.4 wt % GO vs. GnPs	[178] [179]
PCL	Solution processing	12	N/A	12	0.5 wt %	[180]
PLA Mater-bi	Solution Processing and hot pressing	200	N/A	N/A	Effect of few layer graphene vs. GnPs on E_g of both biopolyesters. For comparison other 2D and 3D nanoscale fillers were employed	[65]
Chitosan-tapioca starch	Solution processing	N/A	40	N/A	0.8 wt %	[181]
Regenerated Cellulose	Solution processing	34	56	N/A	3 wt %	[182]
Polyvinyl alcohol	Solution processing	60	40		0.5 wt %	[183]

5.1. Graphene Nanoplatelets Reinforced Polyesters

In particular, polyester bioplastics produced on a large scale, such as poly-lactic acid (PLA), could become more commonplace if their mechanical properties are improved [165]. Indeed, many works deal with the reinforcement of such biopolymers with GnPs [162,164]. For example Botta et al. [166] melt compounded PLA-copolyester biopolymer (BioFlex®) with GnPs, improving the biopolymer's Young or elastic modulus (E_g). They added 1 and 5 wt % GnPs nanofillers into the matrix, increasing E_g of approximately the 40% in the best case. In contrast, the elongation at break and the tensile stress were decreased.

Gonçalves and coauthors [170] melt blended PLA with GnPs. They added different GnP loads (0.1–0.5 wt %), obtaining the maximum mechanical performances (20% increase in tensile stretch, 12% increase of E_g, and 16% increase in toughness) at 0.25 GnPs wt %. They also discovered that the inclusion of GnPs inside PLA did not affect human fibroblasts morphology and metabolic activity at the surface of the samples. These results were obtained with a commercially-available GnPs (XG Science Inc., Lansing, MI, USA, xGnP®, grade C, thickness 10–20 nm, lateral size 1–2 μm). Scaffaro and coworkers modified a PLA matrix with ciprofloxacin and GnPs to obtain biopolymer nanocomposites with antimicrobial properties [171]. GnPs were added at a load of 5 wt %, increasing E_g of the 40%. The elongation at break and tensile strength were almost unchanged. These results can be attributed to the good dispersion level of GnPs achieved during compounding. The incorporation of GnP affected the release of ciprofloxacin without preventing the antimicrobial activity of the obtained materials (see Figure 16). Scaffaro and coauthors [162] obtained similar results with BioFlex® as polymer matrix. Scaffaro and coauthors employed a commercially-available GnPs from XG Science Inc. (grade C750, thickness lower than 2 nm, lateral size 1–2 μm).

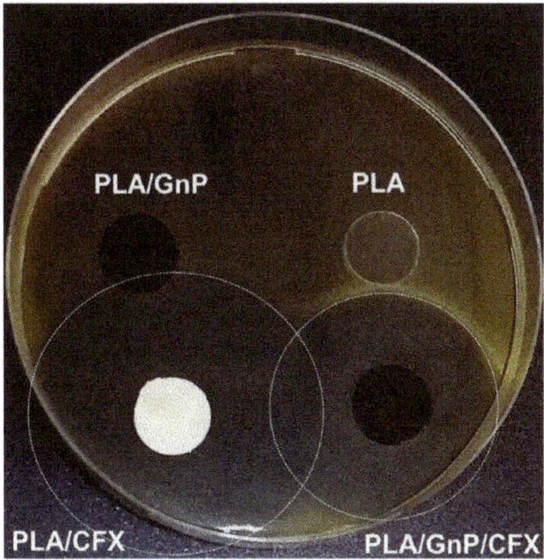

Figure 16. PLA-GnPs-ciprofloxacin composites and their antimicrobial release. Agar diffusion tests were performed with the purpose of investigating if the incorporation of ciprofloxacin inside the PLA specimens was conferring antimicrobial activity to the manufactured films. As bacterium, Scaffaro and coauthors selected *M. luteus*. Both pristine PLA and PLA/GnP displayed no antibacterial properties. In contrast, large bacterial growth inhibition halos were detected near both films, including ciprofloxacin. The presence of GnP led to a reduction of the inhibition zone; therefore, nanoflakes influenced not only the mechanical properties (see text), but also the ciprofloxacin release of the films. Reprinted with permission from Composite part B: Engineering 109, 138–146. Copyright 2017 Elsevier.

Chieng et al. [172] incorporated GnPs inside PLA plasticized with palm oil. With 0.3 wt % nanoflakes melt mixed inside the biopolymer matrix, they were able to increase by approximately 27% and 60% the tensile strength and E_g, respectively. GnPs also decreased the glass transition temperature of the polymer matrix. Similar works always by Chieng et al. were performed mixing PLA, Poly (ethylene glycol), palm oil and GnPs, obtaining similar mechanical improvement [173], and increasing the nanocomposite's antibacterial properties [174]. Other works dealing with melt-mixing inclusion of GnPs inside PLA were published by Narimissa et al. [175,176]. Recently, Gao and coauthors [177] performed a study on the effect of GnP size on the mechanical properties of PLA composites. They used two different commercially available GnPs with lateral sizes of 15 and 1 μm, named large and small, respectively. It was observed that 5 wt % concentrations of the large GnPs augmented E_g by 24%, while the small nanoflakes increased the elastic modulus by 10% at the same concentration. The obtained biocomposites were also electrically conductive, showing a lower percolation threshold for the large GnPs (7 wt % filler loading) compared with the small (13 wt % GnPs concentration).

Other works compare the performance of graphene platelets and graphene oxide inside PLA. For example, Pinto and coauthors [178] reinforced thin films of PLA biopolymer with either GnPs or GO. They used solution processing procedures to fabricate the biopolymers. Both nanoflakes had an optimized load identified at about 0.4 wt % (see Figure 17). In these conditions, GnPs increased E_g by 156% and yield strength by 129%. GO produced similar improvements. Additionally, permeability to nitrogen and oxygen diminished three and four times in films loaded with GO or GnP, respectively. Similar results were achieved by Chieng et al. in this work [179], but by employing melt blending as the production method. In this work [184] Pinto et al. also tested the biocompatibility of reinforced PLA-GnPs and PLA-GO composite materials, finding that low concentrations of graphene-based filler can be safely incorporated, improving their mechanical properties. More details on graphene-based material biocompatibility can be found in another paper [185] by Pinto and coworker.

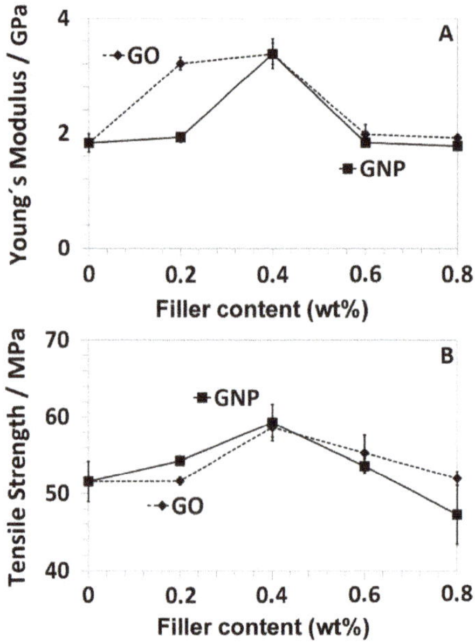

Figure 17. Variation of the Young's Modulus and tensile strength as a function of the nanofiller content of GnPs or GO included inside PLA. Both graphene-based nanofillers increase the mechanical performance of the polymer matrix at low wt % inclusion. Reprinted with permission from Polymer International 62, 1, 33–40. Copyright 2012 Wiley.

Polycaprolactone (PCL), another heavily employed bio-polyester, was coupled with GnPs to enhance its mechanical properties. Wang et al. [180] modified GnPs in water with poly(sodium 4-styrenesulfonate), and compounded such dispersion with PCL. At the best GnP concentration (0.5 wt %), they were able to enhance both the Young modulus and the elongation at a break of approximately the 12%. At 1 wt % GnPs loadings, agglomeration started and the tensile strength and elongation at break reduced considerably. The addition of GnPs augmented notably also the crystallization kinetics. Indeed, only 0.05 wt % nanoflake inclusion triggered a nearly 6 times improvement in crystallization rate.

PLA and Mater-bi® (a blend of PCL and starch) were reinforced with GnPs in our recent work [65]. These two bio-polyesters were mixed with various types of few layer graphene (FLG) and commercially available GnPs. Free standing biocomposites were manufactured by solvent casting and hot-pressing. Exhaustive mechanical measurements were conducted in order to study the effect of FLGs and GnPs thickness and lateral size on the elastic modulus of both polymers. For comparison purposes, other 2D and 3D nanoscale fillers like iron oxides (see Figure 18), clay, and carbon black were used. Under solvent casting conditions (randomly oriented nanoflakes in the polymer matrix), FLG and GnPs did not perform better compared to other model fillers in increasing the elastic modulus of Mater-bi®. On the other hand, both FLGs and large and thick commercially available GnPs increased the elastic moduli of PLA biocomposites more than other 2D and 3D fillers. In the case of hot-pressing induced alignment of the 2D flakes within the polymer matrices, large, many-layer GnPs induced better elastic moduli enhancement compared to FLGs and other 2D and 3D fillers. In particular, GnPs improved the Young Modulus of the Mater-Bi® matrix of the 200%, while PLA enhanced its modulus by 35%. A theoretical model described in the paper is in good agreement with the experimental findings. The highest Young Modulus improvements were achieved with commercially available GnPs from Strem Chemical and Directa Plus S.P.A. For details see this report [65].

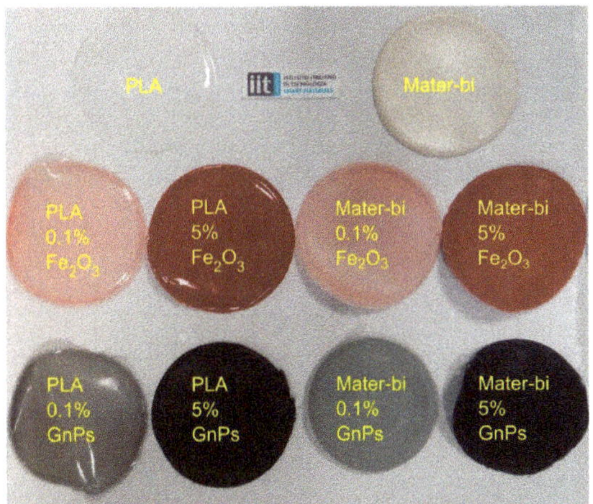

Figure 18. Photograph of different biocomposite films from PLA and Mater-Bi® biopolymers. Whitish and transparent films featured on the top are the pure Mater-bi and PLA matrices. Red films contain Iron Oxides and grey composites include commercially available GnPs. Reprinted with permission from Carbon 109, 331–339. Copyright 2016 Elsevier.

5.2. Reinforcement of Natural Polymer

Apart from bio-polyesters, other biopolymers were also reinforced with GnPs. Ashori et al. [181] included GnPs in chitosan-tapioca starch biocomposite films. The films were produced through a solvent casting method. The best results were achieved incorporating 0.8 wt % of carbon nanofiller, increasing the tensile strength by approximately 40% compared to biocomposites without GnPs. Additionally, the water vapor transmission rate decreased with the addition of a carbon-based nanofiller, while the thermal stability of the nanocomposite increased. Studies on unblended chitosan and starch were also performed by the same authors [186].

Mahmoudian and collaborators [182] prepared regenerated cellulose-GnPs nanocomposites using a solution casting method. At 3 wt % GnP concentration, the elastic modulus improved by 34%, and the tensile strength by 56%. The films also exhibited enhanced carbon dioxide and oxygen gas barrier properties. Gopiraman and collaborators [187] reinforced with GnPs cellulose acetate nanofibers produced by electrospinning. The diameter of the fiber decreased with increased filler amount. At 4 wt % concentrations, the biocomposite showed the highest Young's modulus (approximately 700 MPa), which was about 3.5 times higher than the pristine cellulose acetate fibrous mats. Thayumanavan et al. [183] mixed polyvinyl alcohol, sodium alginate, and GnPs by means of solvent-based techniques. They reported that sodium alginate helps the dispersion of GnPs inside the polymer matrix. By adding 0.5 wt % of nanofiller, they improved the tensile strength and modulus of pure polyvinyl alcohol by approximately 40% and 60% respectively. They obtained similar results by adding a surfactant during the preparation of GnP dispersions [188].

5.3. Reinforced Bioplastics Outlook

In summary, the field of GnPs-reinforced biopolymers is attracting attention. The reinforcement of commercially available bio-polyesters (i.e., PLA) is extensively investigated in light of the possible substitution of oil-derived plastics with bio-based ones. Indeed, today, numerous PLA-GnPs 3D-printable filaments [189] have already been produced by different companies, and are already available in the market. Studies related to the reinforcement of other bio-polyesters such as PCL are still scarce. Indeed, the inclusion of GnPs inside PCL was mainly targeted for cell proliferation and tissue engineering [190,191]. Together with PCL, other biopolymers or natural polymers also reinforced with GnPs, such as starches and cellulosic materials, have good potential for the most demanding mechanical applications, but must be constantly improved. So far, GO was often coupled with such biopolymers, considering its superior chemical bonding with the host matrix; however, since GnPs theoretically have a higher Young's modulus [151], innovative techniques to incorporate GnPs in such biopolymers can pave the way for an enlargement of bioplastic-based structural applications.

6. Conclusions and Outlook

Graphene is foreseen as the breakthrough material of the 21st century. Nevertheless, since there is not a convenient mass-production method so far, other graphene-family products already industrially available are expanding in the market. In particular GnPs are receiving increased interest considering their nano-powder form and appealing chemo-physical properties, which make them a material of choice for advanced nanocomposites. In this review, we mostly concentrated on GnP-based emerging fields, such as flexible and wearable electronics, smart sensing, and reinforced biocomposites.

GnPs-based flexible electronics was thoroughly investigated and appears to have made a significant impact already. Many different approaches were proposed; freestanding GnP films and GnP inclusion in polymers are two examples. However, the most promising results were obtained so far by functionalizing flexible plastic substrates with pure nanoflakes or with polymer-GnPs conductive inks. With such approaches, researchers obtained durable and high-performance devices such as antennas, compliant electrodes for energy applications, and lightweight electromagnetic interference shielding films. Furthermore, extensive use of cellulose substrates, sometimes coupled with biopolymers, has been shown to be promising for the construction of sustainable flexible GnP-based technologies, and for an eco-friendly electronic waste management [192].

Wearable electronics require flexibility, foldability, stretchability, and washability, and at the same time, the ability to maintain a satisfactory electrical conductivity [48]. Nanocomposites, in general, are suitable for conductive wearable technologies [80] due to their intrinsic plastic mechanical properties and ease of manufacturing, and to the large spectra of properties accessible with different nanoparticles. Nonetheless, there is still the need for significant efforts to bring about the commercialization of reliable and washable GnP-based materials for innovative wearable and electrical conductive technologies. New and innovative approaches which bind conductive nanoflakes to textiles, even under severe mechanical stresses and laundry cycles, still need to be implemented.

In contrast, GnP-based smart sensing has already made promising progress. Ad-hoc material combinations that ensure stretchability and tunable electrical features appear to have made a significant impact as new generation wearable sensor technologies. In particular, since graphene-based material has exhibited one of the highest gauge factors ever reported, strain sensing mechanisms have somewhat matured for many uses, ranging from structural and human health monitoring to automotive and sports applications. So far, the combination of PDMS and GnPs were mainly explored, but there are further possibilities considering other elastomeric materials like thermoplastic polyurethanes, rubbers (e.g., nitrile and natural types), and gel-like constructs such as silly putty. Applications such as tactile devices and electronic robotic skins will benefit from the spread-out of GnPs-based smart sensors.

Last but not least, we reviewed the recent advances in GnP-reinforced biopolymer composites. There are already commercial products like skis or tires which benefits from GnP inclusion in oil-derived polymers. Lately, however, the need to reinforce biopolymers also emerged as a strategy to fill the performance gap between traditional long-lasting polymers and bioplastics. Through these strategies, wider use and larger scale use of biopolymers are targeted also for structural applications. PLA is the most investigated bioplastic coupled with GnPs. It has already shown satisfactory results. Indeed, there are already companies marketing GnPs-reinforced PLA 3D printing filaments. In contrast, additional efforts are required for effective GnP inclusion inside other bio-polyesters (like polycaprolactone), starches, and cellulose-based materials. Considering all the results reviewed herein, and potential future developments, we believe that innovative materials and products based on GnPs in polymers will continue evolving towards commercialization and industrialization.

Funding: This research received no external funding.

Conflicts of Interest: The authors declare no conflicts of interest.

References

1. Novoselov, K.S.; Geim, A.K.; Morozov, S.; Jiang, D.; Katsnelson, M.; Grigorieva, I.; Dubonos, S.; Firsov, A.A. Two-dimensional gas of massless dirac fermions in grapheme. *Nature* **2005**, *438*, 197–200. [CrossRef] [PubMed]
2. Geim, A.K. Graphene: Status and prospects. *Science* **2009**, *324*, 1530–1534. [CrossRef] [PubMed]
3. Novoselov, K.S.; Geim, A.K.; Morozov, S.V.; Jiang, D.; Zhang, Y.; Dubonos, S.V.; Grigorieva, I.V.; Firsov, A.A. Electric field effect in atomically thin carbon films. *Science* **2004**, *306*, 666–669. [CrossRef] [PubMed]
4. Stankovich, S.; Dikin, D.A.; Piner, R.D.; Kohlhaas, K.A.; Kleinhammes, A.; Jia, Y.; Wu, Y.; Nguyen, S.T.; Ruoff, R.S. Synthesis of graphene-based nanosheets via chemical reduction of exfoliated graphite oxide. *Carbon* **2007**, *45*, 1558–1565. [CrossRef]
5. Ferrari, A.C.; Bonaccorso, F.; Fal'Ko, V.; Novoselov, K.S.; Roche, S.; Bøggild, P.; Borini, S.; Koppens, F.H.; Palermo, V.; Pugno, N.; et al. Science and technology roadmap for graphene, related two-dimensional crystals, and hybrid systems. *Nanoscale* **2015**, *7*, 4598–4810. [CrossRef] [PubMed]
6. Geim, A.K.; Novoselov, K.S. The rise of grapheme. *Nat. Mater.* **2007**, *6*, 183–191. [CrossRef] [PubMed]
7. Siochi, E.J. Graphene in the sky and beyond. *Nat. Nanotechnol.* **2014**, *9*, 745–747. [CrossRef] [PubMed]
8. Su, C.-Y.; Lu, A.-Y.; Xu, Y.; Chen, F.-R.; Khlobystov, A.N.; Li, L.-J. High-quality thin graphene films from fast electrochemical exfoliation. *ACS Nano* **2011**, *5*, 2332–2339. [CrossRef] [PubMed]
9. Bae, S.; Kim, H.; Lee, Y.; Xu, X.; Park, J.-S.; Zheng, Y.; Balakrishnan, J.; Lei, T.; Kim, H.R.; Song, Y.I.; et al. Roll-to-roll production of 30-inch graphene films for transparent electrodes. *Nat. Nanotechnol.* **2010**, *5*, 574–578. [CrossRef] [PubMed]
10. Novoselov, K.S.; Fal, V.; Colombo, L.; Gellert, P.; Schwab, M.; Kim, K. A roadmap for grapheme. *Nature* **2012**, *490*, 192–200. [CrossRef] [PubMed]
11. Park, S.; Ruoff, R.S. Chemical methods for the production of graphenes. *Nat. Nanotechnol.* **2009**, *4*, 217–224. [CrossRef] [PubMed]
12. Paton, K.R.; Varrla, E.; Backes, C.; Smith, R.J.; Khan, U.; O'Neill, A.; Boland, C.; Lotya, M.; Istrate, O.M.; King, P.; et al. Scalable production of large quantities of defect-free few-layer graphene by shear exfoliation in liquids. *Nat. Mater.* **2014**, *13*, 624–630. [CrossRef] [PubMed]
13. Chen, C.-H.; Yang, S.-W.; Chuang, M.-C.; Woon, W.-Y.; Su, C.-Y. Towards the continuous production of high crystallinity graphene via electrochemical exfoliation with molecular in situ encapsulation. *Nanoscale* **2015**, *7*, 15362–15373. [CrossRef] [PubMed]
14. Li, X.; Cai, W.; An, J.; Kim, S.; Nah, J.; Yang, D.; Piner, R.; Velamakanni, A.; Jung, I.; Tutuc, E.; et al. Large-area synthesis of high-quality and uniform graphene films on copper foils. *Science* **2009**, *324*, 1312–1314. [CrossRef] [PubMed]
15. Hernandez, Y.; Nicolosi, V.; Lotya, M.; Blighe, F.M.; Sun, Z.; De, S.; McGovern, I.; Holland, B.; Byrne, M.; Gun'Ko, Y.K.; et al. High-yield production of graphene by liquid-phase exfoliation of graphite. *Nat. Nanotechnol.* **2008**, *3*, 563–568. [CrossRef] [PubMed]
16. Lotya, M.; Hernandez, Y.; King, P.J.; Smith, R.J.; Nicolosi, V.; Karlsson, L.S.; Blighe, F.M.; De, S.; Wang, Z.; McGovern, I.; et al. Liquid phase production of graphene by exfoliation of graphite in surfactant/water solutions. *J. Am. Chem. Soc.* **2009**, *131*, 3611–3620. [CrossRef] [PubMed]
17. Young, R.J.; Kinloch, I.A.; Gong, L.; Novoselov, K.S. The mechanics of graphene nanocomposites: A review. *Compos. Sci. Technol.* **2012**, *72*, 1459–1476. [CrossRef]
18. Castillo, A.E.D.R.; Pellegrini, V.; Ansaldo, A.; Ricciardella, F.; Sun, H.; Marasco, L.; Buha, J.; Dang, Z.; Gagliani, L.; Lago, E.; et al. High-yield production of 2d crystals by wet-jet milling. *Mater. Horiz.* **2018**. [CrossRef]

19. Wick, P.; Louw-Gaume, A.E.; Kucki, M.; Krug, H.F.; Kostarelos, K.; Fadeel, B.; Dawson, K.A.; Salvati, A.; Vázquez, E.; Ballerini, L.; et al. Classification framework for graphene-based materials. *Angew. Chem. Int. Ed.* **2014**, *53*, 7714–7718. [CrossRef] [PubMed]

20. Jang, B.Z.; Zhamu, A. Processing of nanographene platelets (ngps) and ngp nanocomposites: A review. *J. Mater. Sci.* **2008**, *43*, 5092–5101. [CrossRef]

21. Sengupta, R.; Bhattacharya, M.; Bandyopadhyay, S.; Bhowmick, A.K. A review on the mechanical and electrical properties of graphite and modified graphite reinforced polymer composites. *Prog. Polym. Sci.* **2011**, *36*, 638–670. [CrossRef]

22. Yang, S.-Y.; Lin, W.-N.; Huang, Y.-L.; Tien, H.-W.; Wang, J.-Y.; Ma, C.-C.M.; Li, S.-M.; Wang, Y.-S. Synergetic effects of graphene platelets and carbon nanotubes on the mechanical and thermal properties of epoxy composites. *Carbon* **2011**, *49*, 793–803. [CrossRef]

23. Chung, D. A review of exfoliated graphite. *J. Mater. Sci.* **2016**, *51*, 554–568. [CrossRef]

24. Zhang, M.; Li, Y.; Su, Z.; Wei, G. Recent advances in the synthesis and applications of graphene–polymer nanocomposites. *Polym. Chem.* **2015**, *6*, 6107–6124. [CrossRef]

25. Shen, J.; Hu, Y.; Li, C.; Qin, C.; Ye, M. Synthesis of amphiphilic graphene nanoplatelets. *Small* **2009**, *5*, 82–85. [CrossRef] [PubMed]

26. Masood, M.T.; Papadopoulou, E.L.; Heredia-Guerrero, J.A.; Bayer, I.S.; Athanassiou, A.; Ceseracciu, L. Graphene and polytetrafluoroethylene synergistically improve the tribological properties and adhesion of nylon 66 coatings. *Carbon* **2017**, *123*, 26–33. [CrossRef]

27. Tabandeh-Khorshid, M.; Omrani, E.; Menezes, P.L.; Rohatgi, P.K. Tribological performance of self-lubricating aluminum matrix nanocomposites: Role of graphene nanoplatelets. *Eng. Sci. Technol. Int. J.* **2016**, *19*, 463–469. [CrossRef]

28. Das, A.; Kasaliwal, G.R.; Jurk, R.; Boldt, R.; Fischer, D.; Stöckelhuber, K.W.; Heinrich, G. Rubber composites based on graphene nanoplatelets, expanded graphite, carbon nanotubes and their combination: A comparative study. *Compos. Sci. Technol.* **2012**, *72*, 1961–1967. [CrossRef]

29. Prolongo, S.; Jimenez-Suarez, A.; Moriche, R.; Ureña, A. In situ processing of epoxy composites reinforced with graphene nanoplatelets. *Compos. Sci. Technol.* **2013**, *86*, 185–191. [CrossRef]

30. Rashad, M.; Pan, F.; Tang, A.; Asif, M. Effect of graphene nanoplatelets addition on mechanical properties of pure aluminum using a semi-powder method. *Prog. Natl. Sci. Mater. Int.* **2014**, *24*, 101–108. [CrossRef]

31. Yue, L.; Pircheraghi, G.; Monemian, S.A.; Manas-Zloczower, I. Epoxy composites with carbon nanotubes and graphene nanoplatelets–dispersion and synergy effects. *Carbon* **2014**, *78*, 268–278. [CrossRef]

32. Abbasi, A.; Sadeghi, G.M.M.; Ghasemi, I.; Shahrousvand, M. Shape memory performance of green in situ polymerized nanocomposites based on polyurethane/graphene nanoplatelets: Synthesis, properties, and cell behavior. *Polym. Compos.* **2017**. [CrossRef]

33. Lashgari, S.; Karrabi, M.; Ghasemi, I.; Azizi, H.; Messori, M.; Paderni, K. Shape memory nanocomposite of poly (L-lactic acid)/graphene nanoplatelets triggered by infrared light and thermal heating. *Express Polym. Lett.* **2016**, *10*, 349–359. [CrossRef]

34. Zhang, Z.-X.; Dou, J.-X.; He, J.-H.; Xiao, C.-X.; Shen, L.-Y.; Yang, J.-H.; Wang, Y.; Zhou, Z.-W. Electrically/infrared actuated shape memory composites based on a bio-based polyester blend and graphene nanoplatelets and their excellent self-driven ability. *J. Mater. Chem. C* **2017**, *5*, 4145–4158. [CrossRef]

35. Cui, Y.; Kundalwal, S.; Kumar, S. Gas barrier performance of graphene/polymer nanocomposites. *Carbon* **2016**, *98*, 313–333. [CrossRef]

36. Wu, H.; Drzal, L.T. Graphene nanoplatelet paper as a light-weight composite with excellent electrical and thermal conductivity and good gas barrier properties. *Carbon* **2012**, *50*, 1135–1145. [CrossRef]

37. Dittrich, B.; Wartig, K.-A.; Hofmann, D.; Mülhaupt, R.; Schartel, B. Flame retardancy through carbon nanomaterials: Carbon black, multiwall nanotubes, expanded graphite, multi-layer graphene and graphene in polypropylene. *Polym. Degrad. Stab.* **2013**, *98*, 1495–1505. [CrossRef]

38. Inuwa, I.; Hassan, A.; Wang, D.-Y.; Samsudin, S.; Haafiz, M.M.; Wong, S.; Jawaid, M. Influence of exfoliated graphite nanoplatelets on the flammability and thermal properties of polyethylene terephthalate/polypropylene nanocomposites. *Polym. Degrad. Stab.* **2014**, *110*, 137–148. [CrossRef]

39. Lin, C.; Chung, D. Graphite nanoplatelet pastes vs. carbon black pastes as thermal interface materials. *Carbon* **2009**, *47*, 295–305. [CrossRef]

40. Prolongo, S.; Moriche, R.; Jiménez-Suárez, A.; Sánchez, M.; Ureña, A. Epoxy adhesives modified with graphene for thermal interface materials. *J. Adhes.* **2014**, *90*, 835–847. [CrossRef]

41. Shtein, M.; Nadiv, R.; Buzaglo, M.; Kahil, K.; Regev, O. Thermally conductive graphene-polymer composites: Size, percolation, and synergy effects. *Chem. Mater.* **2015**, *27*, 2100–2106. [CrossRef]

42. Yadav, S.K.; Cho, J.W. Functionalized graphene nanoplatelets for enhanced mechanical and thermal properties of polyurethane nanocomposites. *Appl. Surf. Sci.* **2013**, *266*, 360–367. [CrossRef]

43. Cataldi, P.; Bayer, I.S.; Bonaccorso, F.; Pellegrini, V.; Athanassiou, A.; Cingolani, R. Foldable conductive cellulose fiber networks modified by graphene nanoplatelet-bio-based composites. *Adv. Electron. Mater.* **2015**, *1*. [CrossRef]

44. Jiang, X.; Drzal, L.T. Reduction in percolation threshold of injection molded high-density polyethylene/exfoliated graphene nanoplatelets composites by solid state ball milling and solid state shear pulverization. *J. Appl. Polym. Sci.* **2012**, *124*, 525–535. [CrossRef]

45. Sabzi, M.; Jiang, L.; Liu, F.; Ghasemi, I.; Atai, M. Graphene nanoplatelets as poly (lactic acid) modifier: Linear rheological behavior and electrical conductivity. *J. Mater. Chem. A* **2013**, *1*, 8253–8261. [CrossRef]

46. Yu, A.; Ramesh, P.; Sun, X.; Bekyarova, E.; Itkis, M.E.; Haddon, R.C. Enhanced thermal conductivity in a hybrid graphite nanoplatelet–carbon nanotube filler for epoxy composites. *Adv. Mater.* **2008**, *20*, 4740–4744. [CrossRef]

47. Shahil, K.M.; Balandin, A.A. Graphene–multilayer graphene nanocomposites as highly efficient thermal interface materials. *Nano Lett.* **2012**, *12*, 861–867. [CrossRef] [PubMed]

48. Jang, H.; Park, Y.J.; Chen, X.; Das, T.; Kim, M.-S.; Ahn, J.-H. Graphene-based flexible and stretchable electronics. *Adv. Mater.* **2016**, *28*, 4184–4202. [CrossRef] [PubMed]

49. Cummins, G.; Desmulliez, M.P. Inkjet printing of conductive materials: A review. *Circuit World* **2012**, *38*, 193–213. [CrossRef]

50. Yang, W.; Wang, C. Graphene and the related conductive inks for flexible electronics. *J. Mater. Chem. C* **2016**, *4*, 7193–7207. [CrossRef]

51. King, J.A.; Via, M.D.; Morrison, F.A.; Wiese, K.R.; Beach, E.A.; Cieslinski, M.J.; Bogucki, G.R. Characterization of exfoliated graphite nanoplatelets/polycarbonate composites: Electrical and thermal conductivity, and tensile, flexural, and rheological properties. *J. Compos. Mater.* **2012**, *46*, 1029–1039. [CrossRef]

52. Papadopoulou, E.L.; Pignatelli, F.; Marras, S.; Marini, L.; Davis, A.; Athanassiou, A.; Bayer, I.S. Nylon 6, 6/graphene nanoplatelet composite films obtained from a new solvent. *RSC Adv.* **2016**, *6*, 6823–6831. [CrossRef]

53. Hameed, N.; Dumée, L.F.; Allioux, F.-M.; Reghat, M.; Church, J.S.; Naebe, M.; Magniez, K.; Parameswaranpillai, J.; Fox, B.L. Graphene based room temperature flexible nanocomposites from permanently cross-linked networks. *Sci. Rep.* **2018**, *8*, 2803. [CrossRef] [PubMed]

54. Tian, M.; Huang, Y.; Wang, W.; Li, R.; Liu, P.; Liu, C.; Zhang, Y. Temperature-dependent electrical properties of graphene nanoplatelets film dropped on flexible substrates. *J. Mater. Res.* **2014**, *29*, 1288–1294. [CrossRef]

55. Wróblewski, G.; Janczak, D. Screen printed, transparent, and flexible electrodes based on graphene nanoplatelet pastes. In *Photonics Applications in Astronomy, Communications, Industry, and High-Energy Physics Experiments 2012*; International Society for Optics and Photonics: Bellingham, WA, USA, 2012; Volume 8454, p. 84541E.

56. Seekaew, Y.; Lokavee, S.; Phokharatkul, D.; Wisitsoraat, A.; Kerdcharoen, T.; Wongchoosuk, C. Low-cost and flexible printed graphene–pedot: Pss gas sensor for ammonia detection. *Org. Electron.* **2014**, *15*, 2971–2981. [CrossRef]

57. Huang, X.; Leng, T.; Chang, K.H.; Chen, J.C.; Novoselov, K.S.; Hu, Z. Graphene radio frequency and microwave passive components for low cost wearable electronics. *2D Mater.* **2016**, *3*, 025021. [CrossRef]

58. Mates, J.E.; Bayer, I.S.; Salerno, M.; Carroll, P.J.; Jiang, Z.; Liu, L.; Megaridis, C.M. Durable and flexible graphene composites based on artists' paint for conductive paper applications. *Carbon* **2015**, *87*, 163–174. [CrossRef]

59. Hyun, W.J.; Park, O.O.; Chin, B.D. Foldable graphene electronic circuits based on paper substrates. *Adv. Mater.* **2013**, *25*, 4729–4734. [CrossRef] [PubMed]

60. Scidà, A.; Haque, S.; Treossi, E.; Robinson, A.; Smerzi, S.; Ravesi, S.; Borini, S.; Palermo, V. Application of graphene-based flexible antennas in consumer electronic devices. *Mater. Today* **2018**, *21*, 223–230. [CrossRef]

61. Oh, J.S.; Oh, J.S.; Sung, D.I.; Yeom, G.Y. Fabrication of high-performance graphene nanoplatelet-based transparent electrodes via self-interlayer-exfoliation control. *Nanoscale* **2018**, *10*, 2351–2362. [CrossRef] [PubMed]

62. La Notte, L.; Cataldi, P.; Ceseracciu, L.; Bayer, I.S.; Athanassiou, A.; Marras, S.; Villari, E.; Brunetti, F.; Reale, A. Fully-sprayed flexible polymer solar cells with a cellulose-graphene electrode. *Mater. Today Energy* **2018**, *7*, 105–112. [CrossRef]

63. Cataldi, P.; Bonaccorso, F.; Castillo, A.E.D.; Pellegrini, V.; Jiang, Z.; Liu, L.; Boccardo, N.; Canepa, M.; Cingolani, R.; Athanassiou, A.; et al. Cellulosic graphene biocomposites for versatile high-performance flexible electronic applications. *Adv. Electron. Mater.* **2016**. [CrossRef]

64. Liu, X.; Zou, Q.; Wang, T.; Zhang, L. Electrically conductive graphene-based biodegradable polymer composite films with high thermal stability and flexibility. *Nano* **2018**, *13*, 1850033. [CrossRef]

65. Cataldi, P.; Bayer, I.S.; Nanni, G.; Athanassiou, A.; Bonaccorso, F.; Pellegrini, V.; Castillo, A.E.D.; Ricciardella, F.; Artyukhin, S.; Tronche, M.-A.; et al. Effect of graphene nano-platelet morphology on the elastic modulus of soft and hard biopolymers. *Carbon* **2016**, *109*, 331–339. [CrossRef]

66. Michel, M.; Biswas, C.; Tiwary, C.S.; Saenz, G.A.; Hossain, R.F.; Ajayan, P.; Kaul, A.B. A thermally-invariant, additively manufactured, high-power graphene resistor for flexible electronics. *2D Mater.* **2017**, *4*, 025076. [CrossRef]

67. Zhan, Y.; Lavorgna, M.; Buonocore, G.; Xia, H. Enhancing electrical conductivity of rubber composites by constructing interconnected network of self-assembled graphene with latex mixing. *J. Mater. Chem.* **2012**, *22*, 10464–10468. [CrossRef]

68. Irimia-Vladu, M. "Green" electronics: Biodegradable and biocompatible materials and devices for sustainable future. *Chem. Soc. Rev.* **2014**, *43*, 588–610. [CrossRef] [PubMed]

69. Irimia-Vladu, M.; Gowacki, E.D.; Voss, G.; Bauer, S.; Sariciftci, N.S. Green and biodegradable electronics. *Mater. Today* **2012**, *15*, 340–346. [CrossRef]

70. Stoppa, M.; Chiolerio, A. Wearable electronics and smart textiles: A critical review. *Sensors* **2014**, *14*, 11957–11992. [CrossRef] [PubMed]

71. Bao, Z.; Chen, X. Flexible and stretchable devices. *Adv. Mater.* **2016**, *28*, 4177–4179. [CrossRef] [PubMed]

72. Zeng, W.; Shu, L.; Li, Q.; Chen, S.; Wang, F.; Tao, X.-M. Fiber-based wearable electronics: A review of materials, fabrication, devices, and applications. *Adv. Mater.* **2014**, *26*, 5310–5336. [CrossRef] [PubMed]

73. Babu, K.F.; Dhandapani, P.; Maruthamuthu, S.; Kulandainathan, M.A. One pot synthesis of polypyrrole silver nanocomposite on cotton fabrics for multifunctional property. *Carbohydr. Polym.* **2012**, *90*, 1557–1563. [CrossRef] [PubMed]

74. Ji, X.; Xu, Y.; Zhang, W.; Cui, L.; Liu, J. Review of functionalization, structure and properties of graphene/polymer composite fibers. *Compos. Part A Appl. Sci. Manuf.* **2016**, *87*, 29–45. [CrossRef]

75. Neves, A.I.; Rodrigues, D.P.; Sanctis, A.; Alonso, E.T.; Pereira, M.S.; Amaral, V.S.; Melo, L.V.; Russo, S.; Schrijver, I.; Alves, H.; et al. Towards conductive textiles: Coating polymeric fibres with graphene. *Sci. Rep.* **2017**, *7*, 4250. [CrossRef] [PubMed]

76. Shateri-Khalilabad, M.; Yazdanshenas, M.E. Fabricating electroconductive cotton textiles using graphene. *Carbohydr. Polym.* **2013**, *96*, 190–195. [CrossRef] [PubMed]

77. Zhong, J.; Zhang, Y.; Zhong, Q.; Hu, Q.; Hu, B.; Wang, Z.L.; Zhou, J. Fiber-based generator for wearable electronics and mobile medication. *ACS Nano* **2014**, *8*, 6273–6280. [CrossRef] [PubMed]

78. Liu, W.-W.; Yan, X.-B.; Lang, J.-W.; Peng, C.; Xue, Q.-J. Flexible and conductive nanocomposite electrode based on graphene sheets and cotton cloth for supercapacitor. *J. Mater. Chem.* **2012**, *22*, 17245–17253. [CrossRef]

79. Windmiller, J.R.; Wang, J. Wearable electrochemical sensors and biosensors: A review. *Electroanalysis* **2013**, *25*, 29–46. [CrossRef]

80. Wagner, S.; Bauer, S. Materials for stretchable electronics. *Mrs Bull.* **2012**, *37*, 207–213. [CrossRef]

81. Molina, J. Graphene-based fabrics and their applications: A review. *RSC Adv.* **2016**, *6*, 68261–68291. [CrossRef]
82. Ren, J.; Wang, C.; Zhang, X.; Carey, T.; Chen, K.; Yin, Y.; Torrisi, F. Environmentally-friendly conductive cotton fabric as flexible strain sensor based on hot press reduced graphene oxide. *Carbon* **2017**, *111*, 622–630. [CrossRef]
83. Shateri-Khalilabad, M.; Yazdanshenas, M.E. Preparation of superhydrophobic electroconductive graphene-coated cotton cellulose. *Cellulose* **2013**, *20*, 963–972. [CrossRef]
84. Dong, Z.; Jiang, C.; Cheng, H.; Zhao, Y.; Shi, G.; Jiang, L.; Qu, L. Facile fabrication of light, flexible and multifunctional graphene fibers. *Adv. Mater.* **2012**, *24*, 1856–1861. [CrossRef] [PubMed]
85. Xu, Z.; Gao, C. Graphene fiber: A new trend in carbon fibers. *Mater. Today* **2015**, *18*, 480–492. [CrossRef]
86. Xu, Z.; Sun, H.; Zhao, X.; Gao, C. Ultrastrong fibers assembled from giant graphene oxide sheets. *Adv. Mater.* **2013**, *25*, 188–193. [CrossRef] [PubMed]
87. Neves, A.; Bointon, T.H.; Melo, L.; Russo, S.; de Schrijver, I.; Craciun, M.F.; Alves, H. Transparent conductive graphene textile fibers. *Sci. Rep.* **2015**, *5*, 9866. [CrossRef] [PubMed]
88. Yu, G.; Hu, L.; Vosgueritchian, M.; Wang, H.; Xie, X.; McDonough, J.R.; Cui, X.; Cui, Y.; Bao, Z. Solution-processed graphene/mno2 nanostructured textiles for high-performance electrochemical capacitors. *Nano Lett.* **2011**, *11*, 2905–2911. [CrossRef] [PubMed]
89. Woltornist, S.J.; Alamer, F.A.; McDannald, A.; Jain, M.; Sotzing, G.A.; Adamson, D.H. Preparation of conductive graphene/graphite infused fabrics using an interface trapping method. *Carbon* **2015**, *81*, 38–42. [CrossRef]
90. Sloma, M.; Janczak, D.; Wroblewski, G.; Mlozniak, A.; Jakubowska, M. Electroluminescent structures printed on paper and textile elastic substrates. *Circuit World* **2014**, *40*, 13–16. [CrossRef]
91. Tian, M.; Hu, X.; Qu, L.; Zhu, S.; Sun, Y.; Han, G. Versatile and ductile cotton fabric achieved via layer-by-layer self-assembly by consecutive adsorption of graphene doped pedot: Pss and chitosan. *Carbon* **2016**, *96*, 1166–1174. [CrossRef]
92. Skrzetuska, E.; Puchalski, M.; Krucinska, I. Chemically driven printed textile sensors based on graphene and carbon nanotubes. *Sensors* **2014**, *14*, 16816–16828. [CrossRef] [PubMed]
93. Cataldi, P.; Ceseracciu, L.; Athanassiou, A.; Bayer, I.S. Healable cotton–graphene nanocomposite conductor for wearable electronics. *ACS Appl. Mater. Interfaces* **2017**. [CrossRef] [PubMed]
94. Gong, S.; Cheng, W. One-dimensional nanomaterials for soft electronics. *Adv. Electron. Mater.* **2017**, *3*. [CrossRef]
95. Lu, N.; Kim, D.-H. Flexible and stretchable electronics paving the way for soft robotics. *Soft Robot.* **2014**, *1*, 53–62. [CrossRef]
96. Rogers, J.A.; Someya, T.; Huang, Y. Materials and mechanics for stretchable electronics. *Science* **2010**, *327*, 1603–1607. [CrossRef] [PubMed]
97. Yao, H.-B.; Ge, J.; Wang, C.-F.; Wang, X.; Hu, W.; Zheng, Z.-J.; Ni, Y.; Yu, S.-H. A flexible and highly pressure-sensitive graphene–polyurethane sponge based on fractured microstructure design. *Adv. Mater.* **2013**, *25*, 6692–6698. [CrossRef] [PubMed]
98. Park, M.; Park, J.; Jeong, U. Design of conductive composite elastomers for stretchable electronics. *Nano Today* **2014**, *9*, 244–260. [CrossRef]
99. Yao, S.; Zhu, Y. Nanomaterial-enabled stretchable conductors: Strategies, materials and devices. *Adv. Mater.* **2015**, *27*, 1480–1511. [CrossRef] [PubMed]
100. Yuan, X.; Wei, Y.; Chen, S.; Wang, P.; Liu, L. Bio-based graphene/sodium alginate aerogels for strain sensors. *RSC Adv.* **2016**, *6*, 64056–64064. [CrossRef]
101. Boland, C.S.; Khan, U.; Backes, C.; O'Neill, A.; McCauley, J.; Duane, S.; Shanker, R.; Liu, Y.; Jurewicz, I.; Dalton, A.B.; et al. Sensitive, high-strain, high-rate bodily motion sensors based on graphene–rubber composites. *ACS Nano* **2014**, *8*, 8819–8830. [CrossRef] [PubMed]
102. Jason, N.N.; Wang, S.J.; Bhanushali, S.; Cheng, W. Skin inspired fractal strain sensors using a copper nanowire and graphite microflake hybrid conductive network. *Nanoscale* **2016**, *8*, 16596–16605. [CrossRef] [PubMed]
103. Li, X.; Zhang, R.; Yu, W.; Wang, K.; Wei, J.; Wu, D.; Cao, A.; Li, Z.; Cheng, Y.; Zheng, Q.; et al. Stretchable and highly sensitive graphene-on-polymer strain sensors. *Sci. Rep.* **2012**, *2*, 870. [CrossRef] [PubMed]

104. Hempel, M.; Nezich, D.; Kong, J.; Hofmann, M. A novel class of strain gauges based on layered percolative films of 2d materials. *Nano Lett.* **2012**, *12*, 5714–5718. [CrossRef] [PubMed]

105. Cataldi, P.; Dussoni, S.; Ceseracciu, L.; Maggiali, M.; Natale, L.; Metta, G.; Athanassiou, A.; Bayer, I.S. Carbon nanofiber versus graphene-based stretchable capacitive touch sensors for artificial electronic skin. *Adv. Sci.* **2018**, *5*. [CrossRef] [PubMed]

106. Zahid, M.; Papadopoulou, E.L.; Athanassiou, A.; Bayer, I.S. Strain-responsive mercerized conductive cotton fabrics based on pedot: Pss/graphene. *Mater. Des.* **2017**, *135*, 213–222. [CrossRef]

107. Moriche, R.; Sanchez, M.; Jiménez-Suárez, A.; Prolongo, S.; Urena, A. Strain monitoring mechanisms of sensors based on the addition of graphene nanoplatelets into an epoxy matrix. *Compos. Sci. Technol.* **2016**, *123*, 65–70. [CrossRef]

108. Moriche, R.; Sánchez, M.; Prolongo, S.G.; Jiménez-Suárez, A.; Ureña, A. Reversible phenomena and failure localization in self-monitoring gnp/epoxy nanocomposites. *Compos. Struct.* **2016**, *136*, 101–105. [CrossRef]

109. Wang, B.; Lee, B.-K.; Kwak, M.-J.; Lee, D.-W. Graphene/polydimethylsiloxane nanocomposite strain sensor. *Rev. Sci. Instrum.* **2013**, *84*, 105005. [CrossRef] [PubMed]

110. Boland, C.S.; Khan, U.; Ryan, G.; Barwich, S.; Charifou, R.; Harvey, A.; Backes, C.; Li, Z.; Ferreira, M.S.; Möbius, M.E.; et al. Sensitive electromechanical sensors using viscoelastic graphene-polymer nanocomposites. *Science* **2016**, *354*, 1257–1260. [CrossRef] [PubMed]

111. Cataldi, P.; Ceseracciu, L.; Marras, S.; Athanassiou, A.; Bayer, I.S. Electrical conductivity enhancement in thermoplastic polyurethane-graphene nanoplatelet composites by stretch-release cycles. *Appl. Phys. Lett.* **2017**, *110*, 121904. [CrossRef]

112. Lee, C.; Jug, L.; Meng, E. High strain biocompatible polydimethylsiloxane-based conductive graphene and multiwalled carbon nanotube nanocomposite strain sensors. *Appl. Phys. Lett.* **2013**, *102*, 183511. [CrossRef]

113. Janczak, D.; Soma, M.; Wróblewski, G.; Mozniak, A.; Jakubowska, M. Screen-printed resistive pressure sensors containing graphene nanoplatelets and carbon nanotubes. *Sensors* **2014**, *14*, 17304–17312. [CrossRef] [PubMed]

114. Moriche, R.; Jiménez-Suárez, A.; Sánchez, M.; Prolongo, S.; Ureña, A. Graphene nanoplatelets coated glass fibre fabrics as strain sensors. *Compos. Sci. Technol.* **2017**, *146*, 59–64. [CrossRef]

115. Moriche, R.; Jiménez-Suárez, A.; Sánchez, M.; Prolongo, S.; Ureña, A. Sensitivity, influence of the strain rate and reversibility of gnps based multiscale composite materials for high sensitive strain sensors. *Compos. Sci. Technol.* **2018**, *155*, 100–107. [CrossRef]

116. Park, J.J.; Hyun, W.J.; Mun, S.C.; Park, Y.T.; Park, O.O. Highly stretchable and wearable graphene strain sensors with controllable sensitivity for human motion monitoring. *ACS Appl. Mater. Interfaces* **2015**, *7*, 6317–6324. [CrossRef] [PubMed]

117. Shi, G.; Zhao, Z.; Pai, J.-H.; Lee, I.; Zhang, L.; Stevenson, C.; Ishara, K.; Zhang, R.; Zhu, H.; Ma, J. Highly sensitive, wearable, durable strain sensors and stretchable conductors using graphene/silicon rubber composites. *Adv. Funct. Mater.* **2016**, *26*, 7614–7625. [CrossRef]

118. Zhao, H.; Bai, J. Highly sensitive piezo-resistive graphite nanoplatelet–carbon nanotube hybrids/polydimethylsilicone composites with improved conductive network construction. *ACS Appl. Mater. Interfaces* **2015**, *7*, 9652–9659. [CrossRef] [PubMed]

119. Filippidou, M.; Tegou, E.; Tsouti, V.; Chatzandroulis, S. A flexible strain sensor made of graphene nanoplatelets/polydimethylsiloxane nanocomposite. *Microelectron. Eng.* **2015**, *142*, 7–11. [CrossRef]

120. Rinaldi, A.; Tamburrano, A.; Fortunato, M.; Sarto, M.S. A flexible and highly sensitive pressure sensor based on a pdms foam coated with graphene nanoplatelets. *Sensors* **2016**, *16*, 2148. [CrossRef] [PubMed]

121. Moriche, R.; Prolongo, S.G.; Sánchez, M.; Jiménez-Suárez, A.; Campo, M.; Ureña, A. Strain sensing based on multiscale composite materials reinforced with graphene nanoplatelets. *J. Vis. Exp. JoVE* **2016**. [CrossRef] [PubMed]

122. Yan, C.; Wang, J.; Kang, W.; Cui, M.; Wang, X.; Foo, C.Y.; Chee, K.J.; Lee, P.S. Highly stretchable piezoresistive graphene–nanocellulose nanopaper for strain sensors. *Adv. Mater.* **2014**, *26*, 2022–2027. [CrossRef] [PubMed]

123. Kim, Y.-J.; Cha, J.Y.; Ham, H.; Huh, H.; So, D.-S.; Kang, I. Preparation of piezoresistive nano smart hybrid material based on graphene. *Curr. Appl. Phys.* **2011**, *11*, S350–S352. [CrossRef]

124. Tadakaluru, S.; Thongsuwan, W.; Singjai, P. Stretchable and flexible high-strain sensors made using carbon nanotubes and graphite films on natural rubber. *Sensors* **2014**, *14*, 868–876. [CrossRef] [PubMed]

125. Amjadi, M.; Pichitpajongkit, A.; Lee, S.; Ryu, S.; Park, I. Highly stretchable and sensitive strain sensor based on silver nanowire–elastomer nanocomposite. *ACS Nano* **2014**, *8*, 5154–5163. [CrossRef] [PubMed]

126. Jeong, Y.R.; Park, H.; Jin, S.W.; Hong, S.Y.; Lee, S.-S.; Ha, J.S. Highly stretchable and sensitive strain sensors using fragmentized graphene foam. *Adv. Funct. Mater.* **2015**, *25*, 4228–4236. [CrossRef]

127. Lee, J.; Kim, S.; Lee, J.; Yang, D.; Park, B.C.; Ryu, S.; Park, I. A stretchable strain sensor based on a metal nanoparticle thin film for human motion detection. *Nanoscale* **2014**, *6*, 11932–11939. [CrossRef] [PubMed]

128. Lötters, J.C.; Olthuis, W.; Veltink, P.; Bergveld, P. The mechanical properties of the rubber elastic polymer polydimethylsiloxane for sensor applications. *J. Micromechan. Microeng.* **1997**, *7*, 145–147. [CrossRef]

129. Lu, N.; Lu, C.; Yang, S.; Rogers, J. Highly sensitive skin-mountable strain gauges based entirely on elastomers. *Adv. Funct. Mater.* **2012**, *22*, 4044–4050. [CrossRef]

130. Yamada, T.; Hayamizu, Y.; Yamamoto, Y.; Yomogida, Y.; Izadi-Najafabadi, A.; Futaba, D.N.; Hata, K. A stretchable carbon nanotube strain sensor for human-motion detection. *Nat. Nanotechnol.* **2011**, *6*, 296–301. [CrossRef] [PubMed]

131. Lee, S.W.; Park, J.J.; Park, B.H.; Mun, S.C.; Park, Y.T.; Liao, K.; Seo, T.S.; Hyun, W.J.; Park, O.O. Enhanced sensitivity of patterned graphene strain sensors used for monitoring subtle human body motions. *ACS Appl. Mater. Interfaces* **2017**, *9*, 11176–11183. [CrossRef] [PubMed]

132. Liu, H.; Li, Y.; Dai, K.; Zheng, G.; Liu, C.; Shen, C.; Yan, X.; Guo, J.; Guo, Z. Electrically conductive thermoplastic elastomer nanocomposites at ultralow graphene loading levels for strain sensor applications. *J. Mater. Chem. C* **2016**, *4*, 157–166. [CrossRef]

133. Cheng, Y.; Wang, R.; Sun, J.; Gao, L. A stretchable and highly sensitive graphene-based fiber for sensing tensile strain, bending, and torsion. *Adv. Mater.* **2015**, *27*, 7365–7371. [CrossRef] [PubMed]

134. Rams, J.; Sanchez, M.; Urena, A.; Jimenez-Suarez, A.; Campo, M.; Güemes, A. Use of carbon nanotubes for strain and damage sensing of epoxy-based composites. *Int. J. Smart Nano Mater.* **2012**, *3*, 152–161. [CrossRef]

135. Schmitz, A.; Maiolino, P.; Maggiali, M.; Natale, L.; Cannata, G.; Metta, G. Methods and technologies for the implementation of large-scale robot tactile sensors. *IEEE Trans. Robot.* **2011**, *27*, 389–400. [CrossRef]

136. Zhang, N.; Luan, P.; Zhou, W.; Zhang, Q.; Cai, L.; Zhang, X.; Zhou, W.; Fan, Q.; Yang, F.; Zhao, D.; et al. Highly stretchable pseudocapacitors based on buckled reticulate hybrid electrodes. *Nano Res.* **2014**, *7*, 1680–1690. [CrossRef]

137. Das, T.K.; Prusty, S. Graphene-based polymer composites and their applications. *Polym.-Plast. Technol. Eng.* **2013**, *52*, 319–331. [CrossRef]

138. Kim, H.; Abdala, A.A.; Macosko, C.W. Graphene/polymer nanocomposites. *Macromolecules* **2010**, *43*, 6515–6530. [CrossRef]

139. Kuilla, T.; Bhadra, S.; Yao, D.; Kim, N.H.; Bose, S.; Lee, J.H. Recent advances in graphene based polymer composites. *Prog. Polym. Sci.* **2010**, *35*, 1350–1375. [CrossRef]

140. Mukhopadhyay, P.; Gupta, R.K. Trends and frontiers in graphene-based polymer nanocomposites. *Plast. Eng.* **2011**, *67*, 32–42.

141. Rafiee, M.A. Graphene-Based Composite Materials. Ph.D. Thesis, Rensselaer Polytechnic Institute, Troy, NY, USA, 2011.

142. Hu, K.; Kulkarni, D.D.; Choi, I.; Tsukruk, V.V. Graphene-polymer nanocomposites for structural and functional applications. *Prog. Polym. Sci.* **2014**, *39*, 1934–1972. [CrossRef]

143. Karevan, M.; Kalaitzidou, K. Understanding the property enhancement mechanism in exfoliated graphite nanoplatelets reinforced polymer nanocomposites. *Compos. Interfaces* **2013**, *20*, 255–268. [CrossRef]

144. Zaman, I.; Manshoor, B.; Khalid, A.; Araby, S. From clay to graphene for polymer nanocomposites: A survey. *J. Polym. Res.* **2014**, *21*, 429. [CrossRef]

145. Dimov, D.; Amit, I.; Gorrie, O.; Barnes, M.D.; Townsend, N.J.; Neves, A.I.; Withers, F.; Russo, S.; Craciun, M.F. Ultrahigh performance nanoengineered graphene–concrete composites for multifunctional applications. *Adv. Funct. Mater.* **2018**, 1705183. [CrossRef]

146. Böhm, S. Graphene against corrosion. *Nat. Nanotechnol.* **2014**, *9*, 741–742. [CrossRef] [PubMed]

147. Elmarakbi, A.; Azoti, W. Novel composite materials for automotive applications: Concepts and challenges for energy-efficient and safe vehicles. In Proceedings of the 10th International Conference on Composite Science and Technology, Lisbon, Portugal, 2–4 September 2015.
148. Rafiee, M.; Rafiee, J.; Yu, Z.-Z.; Koratkar, N. Buckling resistant graphene nanocomposites. *Appl. Phys. Lett.* **2009**, *95*, 223103. [CrossRef]
149. Das, D.; Swain, P.; Sahoo, S. Graphene in turbine blades. *Modern Phys. Lett. B* **2016**, *30*, 1650262. [CrossRef]
150. Yavari, F.; Rafiee, M.; Rafiee, J.; Yu, Z.-Z.; Koratkar, N. Dramatic increase in fatigue life in hierarchical graphene composites. *ACS Appl. Mater. Interfaces* **2010**, *2*, 2738–2743. [CrossRef] [PubMed]
151. Ma, J.; Meng, Q.; Zaman, I.; Zhu, S.; Michelmore, A.; Kawashima, N.; Wang, C.H.; Kuan, H.-C. Development of polymer composites using modified, high-structural integrity graphene platelets. *Compos. Sci. Technol.* **2014**, *91*, 82–90. [CrossRef]
152. Rafiee, M.A.; Rafiee, J.; Wang, Z.; Song, H.; Yu, Z.-Z.; Koratkar, N. Enhanced mechanical properties of nanocomposites at low graphene content. *ACS Nano* **2009**, *3*, 3884–3890. [CrossRef] [PubMed]
153. Mittal, G.; Dhand, V.; Rhee, K.Y.; Park, S.-J.; Lee, W.R. A review on carbon nanotubes and graphene as fillers in reinforced polymer nanocomposites. *J. Ind. Eng. Chem.* **2015**, *21*, 11–25. [CrossRef]
154. Boothroyd, S.C.; Johnson, D.W.; Weir, M.P.; Reynolds, C.D.; Hart, J.M.; Smith, A.J.; Clarke, N.; Thompson, R.L.; Coleman, K.S. Controlled structure evolution of graphene networks in polymer composites. *Chem. Mater.* **2018**, *30*, 1524–1530. [CrossRef]
155. Sun, X.; Sun, H.; Li, H.; Peng, H. Developing polymer composite materials: Carbon nanotubes or graphene? *Adv. Mater.* **2013**, *25*, 5153–5176. [CrossRef] [PubMed]
156. Tang, L.-C.; Wan, Y.-J.; Yan, D.; Pei, Y.-B.; Zhao, L.; Li, Y.-B.; Wu, L.-B.; Jiang, J.-X.; Lai, G.-Q. The effect of graphene dispersion on the mechanical properties of graphene/epoxy composites. *Carbon* **2013**, *60*, 16–27. [CrossRef]
157. Moriche, R.; Prolongo, S.; Sánchez, M.; Jiménez-Suárez, A.; Sayagués, M.; Ureña, A. Morphological changes on graphene nanoplatelets induced during dispersion into an epoxy resin by different methods. *Compos. Part B Eng.* **2015**, *72*, 199–205. [CrossRef]
158. Prolongo, S.; Jiménez-Suárez, A.; Moriche, R.; Ureña, A. Graphene nanoplatelets thickness and lateral size influence on the morphology and behavior of epoxy composites. *Eur. Polym. J.* **2014**, *53*, 292–301. [CrossRef]
159. Papageorgiou, D.G.; Kinloch, I.A.; Young, R.J. Mechanical properties of graphene and graphene-based nanocomposites. *Prog. Mater. Sci.* **2017**, *90*, 75–127. [CrossRef]
160. Bayer, I.S. Thermomechanical properties of polylactic acid-graphene composites: A state-of-the-art review for biomedical applications. *Materials* **2017**, *10*, 748. [CrossRef] [PubMed]
161. Heredia-Guerrero, J.A.; Bentez, J.J.; Cataldi, P.; Paul, U.C.; Contardi, M.; Cingolani, R.; Bayer, I.S.; Heredia, A.; Athanassiou, A. All-natural sustainable packaging materials inspired by plant cuticles. *Adv. Sustain. Syst.* **2017**, *1*. [CrossRef]
162. Scaffaro, R.; Botta, L.; Maio, A.; Mistretta, M.C.; la Mantia, F.P. Effect of graphene nanoplatelets on the physical and antimicrobial properties of biopolymer-based nanocomposites. *Materials* **2016**, *9*, 351. [CrossRef] [PubMed]
163. Bordes, P.; Pollet, E.; Avérous, L. Nano-biocomposites: Biodegradable polyester/nanoclay systems. *Prog. Polym. Sci.* **2009**, *34*, 125–155. [CrossRef]
164. Raquez, J.-M.; Habibi, Y.; Murariu, M.; Dubois, P. Polylactide (pla)-based nanocomposites. *Prog. Polym. Sci.* **2013**, *38*, 1504–1542. [CrossRef]
165. Sriprachuabwong, C.; Duangsripat, S.; Sajjaanantakul, K.; Wisitsoraat, A.; Tuantranont, A. Electrolytically exfoliated graphene–polylactide-based bioplastic with high elastic performance. *J. Appl. Polym. Sci.* **2015**, *132*. [CrossRef]
166. Botta, L.; Scaffaro, R.; Mistretta, M.; la Mantia, F. Biopolymer based nanocomposites reinforced with graphene nanoplatelets. *AIP Conf. Proc.* **2016**, *1736*, 020156.
167. Rouf, T.B.; Kokini, J.L. Biodegradable biopolymer–graphene nanocomposites. *J. Mater. Sci.* **2016**, *51*, 9915–9945. [CrossRef]
168. Mittal, V.; Chaudhry, A.U.; Luckachan, G.E. Biopolymer–thermally reduced graphene nanocomposites: Structural characterization and properties. *Mater. Chem. Phys.* **2014**, *147*, 319–332. [CrossRef]

169. Ioniță, M.; Vlăsceanu, G.M.; Watzlawek, A.A.; Voicu, S.I.; Burns, J.S.; Iovu, H. Graphene and functionalized graphene: Extraordinary prospects for nanobiocomposite materials. *Compos. Part B Eng.* **2017**, *121*, 34–57. [CrossRef]
170. Gonçalves, C.; Pinto, A.; Machado, A.V.; Moreira, J.; Gonçalves, I.C.; Magalhães, F. Biocompatible reinforcement of poly (lactic acid) with graphene nanoplatelets. *Polym. Compos.* **2016**. [CrossRef]
171. Scaffaro, R.; Botta, L.; Maio, A.; Gallo, G. Pla graphene nanoplatelets nanocomposites: Physical properties and release kinetics of an antimicrobial agent. *Compos. Part B Eng.* **2017**, *109*, 138–146. [CrossRef]
172. Chieng, B.W.; Ibrahim, N.A.; Yunus, W.M.Z.W.; Hussein, M.Z.; Loo, Y.Y. Effect of graphene nanoplatelets as nanofiller in plasticized poly (lactic acid) nanocomposites. *J. Ther. Anal. Calorim.* **2014**, *118*, 1551–1559. [CrossRef]
173. Chieng, B.W.; Ibrahim, N.A.; Yunus, W.M.Z.W.; Hussein, M.Z. Poly (lactic acid)/poly (ethylene glycol) polymer nanocomposites: Effects of graphene nanoplatelets. *Polymers* **2013**, *6*, 93–104. [CrossRef]
174. Chieng, B.W.; Ibrahim, N.A.; Yunus, W.M.Z.W.; Hussein, M.Z.; Then, Y.Y.; Loo, Y.Y. Reinforcement of graphene nanoplatelets on plasticized poly (lactic acid) nanocomposites: Mechanical, thermal, morphology, and antibacterial properties. *J. Appl. Polym. Sci.* **2015**, *132*. [CrossRef]
175. Narimissa, E.; Gupta, R.K.; Choi, H.J.; Kao, N.; Jollands, M. Morphological, mechanical, and thermal characterization of biopolymer composites based on polylactide and nanographite platelets. *Polym. Compos.* **2012**, *33*, 1505–1515. [CrossRef]
176. Narimissa, E.; Gupta, R.K.; Kao, N.; Choi, H.J.; Jollands, M.; Bhattacharya, S.N. Melt rheological investigation of polylactide-nanographite platelets biopolymer composites. *Polym. Eng. Sci.* **2014**, *54*, 175–188. [CrossRef]
177. Gao, Y.; Picot, O.T.; Bilotti, E.; Peijs, T. Influence of filler size on the properties of poly (lactic acid)(pla)/graphene nanoplatelet (gnp) nanocomposites. *Eur. Polym. J.* **2017**, *86*, 117–131. [CrossRef]
178. Pinto, A.M.; Cabral, J.; Tanaka, D.A.P.; Mendes, A.M.; Magalhães, F.D. Effect of incorporation of graphene oxide and graphene nanoplatelets on mechanical and gas permeability properties of poly (lactic acid) films. *Polym. Int.* **2013**, *62*, 33–40. [CrossRef]
179. Chieng, B.W.; Ibrahim, N.A.; Yunus, W.M.Z.W.; Hussein, M.Z.; Then, Y.Y.; Loo, Y.Y. Effects of graphene nanoplatelets and reduced graphene oxide on poly (lactic acid) and plasticized poly (lactic acid): A comparative study. *Polymers* **2014**, *6*, 2232–2246. [CrossRef]
180. Wang, M.; Deng, X.-Y.; Du, A.-K.; Zhao, T.-H.; Zeng, J.-B. Poly (sodium 4-styrenesulfonate) modified graphene for reinforced biodegradable poly (-caprolactone) nanocomposites. *RSC Adv.* **2015**, *5*, 73146–73154. [CrossRef]
181. Ashori, A.; Bahrami, R. Modification of physico-mechanical properties of chitosan-tapioca starch blend films using nano graphene. *Polym. Plast. Technol. Eng.* **2014**, *53*, 312–318. [CrossRef]
182. Mahmoudian, S.; Wahit, M.U.; Imran, M.; Ismail, A.; Balakrishnan, H. A facile approach to prepare regenerated cellulose/graphene nanoplatelets nanocomposite using room-temperature ionic liquid. *J. Nanosci. Nanotechnol.* **2012**, *12*, 5233–5239. [CrossRef] [PubMed]
183. Thayumanavan, N.; Tambe, P.; Joshi, G.; Shukla, M. Effect of sodium alginate modification of graphene (by "anion-π" type of interaction) on the mechanical and thermal properties of polyvinyl alcohol (pva) nanocomposites. *Compos. Interfaces* **2014**, *21*, 487–506. [CrossRef]
184. Pinto, A.M.; Moreira, S.; Gonçalves, I.C.; Gama, F.M.; Mendes, A.M.; Magalhães, F.D. Biocompatibility of poly (lactic acid) with incorporated graphene-based materials. *Colloids Surf. B Biointerfaces* **2013**, *104*, 229–238. [CrossRef] [PubMed]
185. Pinto, A.M.; Goncalves, I.C.; Magalhães, F.D. Graphene-based materials biocompatibility: A review. *Colloids Surf. B Biointerfaces* **2013**, *111*, 188–202. [CrossRef] [PubMed]
186. Ashori, A. Effects of graphene on the behavior of chitosan and starch nanocomposite films. *Polym. Eng. Sci.* **2014**, *54*, 2258–2263. [CrossRef]
187. Gopiraman, M.; Fujimori, K.; Zeeshan, K.; Kim, B.; Kim, I. Structural and mechanical properties of cellulose acetate/graphene hybrid nanofibers: Spectroscopic investigations. *Express Polym. Lett.* **2013**, *7*, 554–564. [CrossRef]
188. Thayumanavan, N.; Tambe, P.; Joshi, G. Effect of surfactant and sodium alginate modification of graphene on the mechanical and thermal properties of polyvinyl alcohol (pva) nanocomposites. *Cell. Chem. Technol.* **2015**, *49*, 69–80.

189. Bustillos, J.; Montero, D.; Nautiyal, P.; Loganathan, A.; Boesl, B.; Agarwal, A. Integration of graphene in poly (lactic) acid by 3d printing to develop creep and wear-resistant hierarchical nanocomposites. *Polym. Compos.* **2017**. [CrossRef]
190. Qian, Y.; Zhao, X.; Han, Q.; Chen, W.; Li, H.; Yuan, W. An integrated multi-layer 3d-fabrication of pda/rgd coated graphene loaded pcl nanoscaffold for peripheral nerve restoration. *Nat. Commun.* **2018**, *9*, 323. [CrossRef] [PubMed]
191. Wang, W.; Caetano, G.; Ambler, W.S.; Blaker, J.J.; Frade, M.A.; Mandal, P.; Diver, C.; Bártolo, P. Enhancing the hydrophilicity and cell attachment of 3d printed pcl/graphene scaffolds for bone tissue engineering. *Materials* **2016**, *9*, 992. [CrossRef] [PubMed]
192. Cataldi, P.; Heredia-Guerrero, J.A.; Guzman-Puyol, S.; Ceseracciu, L.; La Notte, L.; Reale, A.; Ren, J.; Zhang, Y.; Liu, L.; Miscuglio, M.; et al. Sustainable Electronics Based on Crop Plant Extracts and Graphene: A "Bioadvantaged" Approach. *Adv. Sustain. Syst.* **2018**, 1800069. [CrossRef]

applied sciences

Article

Terahertz Time-Domain Spectroscopy of Graphene Nanoflakes Embedded in Polymer Matrix

Anton Koroliov [1,†], Genyu Chen [2], Kenneth M. Goodfellow [3], A. Nick Vamivakas [3], Zygmunt Staniszewski [4], Peter Sobolewski [4], Mirosława El Fray [4,*], Adam Łaszcz [5], Andrzej Czerwinski [5], Christiaan P. Richter [6] and Roman Sobolewski [1,2,5,*]

[1] Department of Electrical and Computer Engineering and the Laboratory for Laser Energetics, University of Rochester, Rochester, NY 14627-0231, USA; anton.koroliov@ftmc.lt

[2] Materials Science Program and the Laboratory for Laser Energetics, University of Rochester, Rochester, NY 14627, USA; gchen21@ur.rochester.edu

[3] The Institute of Optics, University of Rochester, Rochester, NY 14627, USA; kgoodfel@ur.rochester.edu (K.M.G.); nick.vamivakas@rochester.edu (A.N.V.)

[4] Polymer Institute, West Pomeranian University of Technology, PL-70322 Szczecin, Poland; zstaniszewski@zut.edu.pl (Z.S.); psobolewski@zut.edu.pl (P.S.)

[5] Institute of Electron Technology, PL-02668 Warszawa, Poland; laszcz@ite.waw.pl (A.Ł.); aczerwin@ite.waw.pl (A.C.)

[6] Chemical Engineering, University of Iceland, IS-107 Reykjavik, Iceland; cpr@hi.is

[*] Correspondence: mirfray@zut.edu.pl (M.E.F.); roman.sobolewski@rochester.edu (R.S.); Tel.: +48-91-449-4828 (M.E.F.); +1-585-275-1551 (R.S.)

[†] Current address: Center for Physical Sciences and Technology, LT-10257 Vilnius, Lithuania.

Received: 29 November 2018; Accepted: 14 January 2019; Published: 23 January 2019

Featured Application: We describe the use of terahertz (THz) time-domain spectroscopy to characterize the dispersion of graphene nanoflakes, within a multiblock copolyester matrix. This method probes the dielectric properties of the sample in an almost cm2 cross-section beam path, thereby providing "global" information regarding the dispersion of graphene and its in-situ electronic quality. Because it is a non-destructive testing method, THz time-domain spectroscopy holds great potential for monitoring any nanofiller dispersion in a polymer matrix throughout the product development chain: After polymer nanocomposite synthesis, following processing into a given prototype, and even after product testing.

Abstract: The terahertz time-domain spectroscopy (THz-TDS) technique has been used to obtain transmission THz-radiation spectra of polymer nanocomposites containing a controlled amount of exfoliated graphene. Graphene nanocomposites (1 wt%) that were used in this work were based on poly(ethylene terephthalate-ethylene dilinoleate) (PET-DLA) matrix and were prepared via a kilo-scale (suitable for research and development, and prototyping) in-situ polymerization. This was followed by compression molding into 0.3-mm-thick and 0.9-mm-thick foils. Transmission electron microscopy (TEM) and Raman studies were used to confirm that the graphene nanoflakes dispersed in a polymer matrix consisted of a few-layer graphene. The THz-radiation transients were generated and detected using a low-temperature–grown GaAs photoconductive emitter and detector, both excited by 100-fs-wide, 800-nm-wavelength optical pulses, generated at a 76-MHz repetition rate by a Ti:Sapphire laser. Time-domain signals transmitted through the nitrogen, neat polymer reference, and 1-wt% graphene-polymer nanocomposite samples were recorded and subsequently converted into the spectral domain by means of a fast Fourier transformation. The spectral range of our spectrometer was up to 4 THz, and measurements were taken at room temperature in a dry nitrogen environment. We collected a family of spectra and, based on Fresnel equations, performed a numerical analysis, that allowed us to extract the THz-frequency-range refractive index and absorption coefficient and their dependences on the sample composition and graphene content. Using the Clausius-Mossotti relation, we also managed to estimate the graphene effective dielectric constant to be equal to ~7 $\pm$ 2. Finally, we extracted from our experimental data complex conductivity spectra of graphene nanocomposites and successfully fitted them to the Drude-Smith model, demonstrating that our graphene nanoflakes were isolated in their polymer matrix and exhibited highly localized electron backscattering with a femtosecond relaxation time. Our results shed new light on how the incorporation of exfoliated graphene nanoflakes modifies polymer electrical properties in the THz-frequency range. Importantly, they demonstrate that the complex conductivity analysis is a very efficient, macroscopic and non-destructive (contrary to TEM) tool for the characterization of the dispersion of a graphene nanofiller within a copolyester matrix.

Keywords: graphene; graphene nanoflakes; graphene-polymer nanocomposites; multiblock copolyesters; terahertz time-domain spectroscopy; Drude–Smith model for complex conductivity

---

1. Introduction

Polymer nanocomposites are prepared by including a nanofiller (carbon, ceramic, metal/metal oxide, and/or others) in a polymer matrix in order to alter (or improve) its properties, such as mechanical, electrical, and optical properties, barrier properties, flame resistance, etc. [1]. Recently, a great deal of interest has been focused on carbon nanofillers and, in particular, graphene. Graphene is a two-dimensional (2-D) nanomaterial consisting of sheets of carbon atoms bonded by sp^2 bonds in a hexagonal configuration: Its unique mechanical, electrical, thermal, and optical properties have been extensively studied [2]. These properties have been leveraged in the development of a wide range of different graphene-polymer nanocomposites [3] using a variety of fabrication methods and polymer types for numerous structural and functional applications [4], including, e.g., biomedical devices [5], biosensing [6], and gas barrier membranes [7].

The key distinguishing feature of nano-fillers, as compared with microscale or conventional fillers, is the very high surface-to-volume ratio inherent in their nanoscale size. Therefore, nano-fillers can have an increased interfacial interaction with the polymer matrix, despite being present at very low concentrations (<5 wt%). In this context, along with the size, the dispersion of the nanofiller in the matrix plays a central role in governing the resulting nanocomposite properties [8]. Generally, it is advantageous for a nanofiller to be well dispersed within the polymer matrix; however, in the case of carbon nanofillers, graphene, and graphite nanoplatelets in particular, obtaining high-quality dispersions without aggregation is a major challenge and limitation [4]. Further, control over the dispersion—to obtain ordered or hierarchical organization of the nanofiller within the polymer matrix—holds breakthrough potential [8].

At present, electron microscopy—transmission electron microscopy (TEM) in particular—remains the gold standard for assessing the dispersion of a nanofiller, because it gives a direct nanofiller image of the dispersion and morphology and can offer density quantification [9]. This approach is limited, however, by the fact that it presents only very local information from select fields of view that are always 2-D. As a result, other methods are being investigated, such as x-ray diffraction and small-angle scattering [10], modulated temperature differential scanning calorimetry [11], ultrasonic dynamic mechanical analysis [12], or dynamic rheological analysis [13]. While each of these methods has distinct advantages, none of them can be used on an object or component of a device because they all require dedicated sample preparation and are often destructive. As a result, these methods cannot be used to assess a nanocomposite after it has been processed into its application state without alteration. However, polymer processing methods—and especially industrially and economically viable methods, such as injection or compression molding—have a definite impact on the properties of a given nanocomposite [14], including the dispersion of a nanofiller.

Terahertz time-domain spectroscopy (THz-TDS) offers an alternative as a non-destructive method suitable for testing even relatively large specimens. THz-TDS has already demonstrated its potential for materials characterization [15], including non-destructive testing (NDT) of both polymers [16] and other composite materials [17]. For example, it has been used to assess damage or detect defects in aeronautic components [18,19]. Further, several groups have proposed using this approach to analyze carbon nanotube and/or graphene-polymer nanocomposites to assess the level of dispersion of the nanofiller [20–24].

Here, we report the use of a THz-TDS technique to characterize graphene nanofiller dispersion within multi-block copolyester nanocomposite materials. Both the copolymer matrix and the nanocomposites were developed within the scope of the ElastoKard project in partnership with the Professor Zbigniew Religa Foundation of Cardiac Surgery Development (FRK) to serve as construction materials for extracorporeal heart assist devices in the context of the Polish Artificial Heart Program [25]. We have recently reported the kiloscale (suitable for Research and Development and prototyping) in-situ polymerization synthesis of these materials and their full physicochemical, thermal, and mechanical characterization [26,27]. These co-polymers and nanocomposites have been successfully used to produce prototype extracorporeal heart assist devices (Figure 1a), including the important hemispherical pneumatic membrane composed of the graphene nanocomposite (Figure 1b). Here, the test samples used were processed using the same compression molding process as the prototype pneumatic membrane. To our knowledge, this is the first time that the dispersion of graphene nanoplatelets within a nanocomposite—prepared at an industrially relevant scale and using an economically viable process—has been studied by the THz-TDS method. Consequently, this work serves as important proof of the concept of THz-TDS of nanocomposites within the product development chain. The level of dispersion of graphene nanoflakes within our nanocomposites was assessed based on the THz-range complex conductivity data, fitted using the Drude–Smith model. This THz characterization approach measures and utilizes both the real and the imaginary parts of the dielectric function (or, equivalently, the index of refraction) for both the neat copolymers and graphene nanocomposites, as well as the amount of THz radiation absorbed as the THz probe beam propagates through the sample.

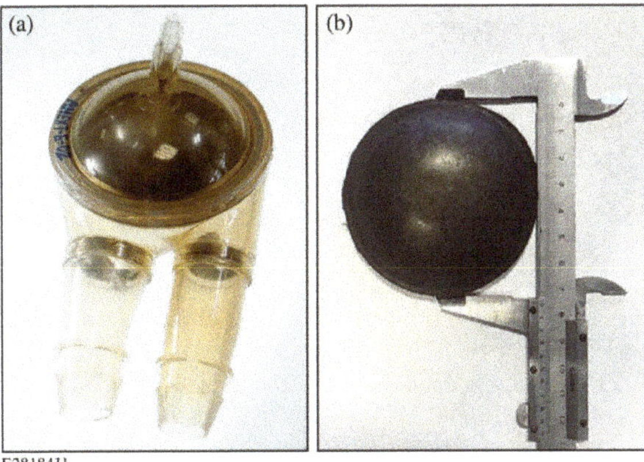

Figure 1. (**a**) Prototype of extracorporeal heart assist device (reprinted with permission [27], copyright 2018, Elsevier); (**b**) graphene nanocomposite pneumatic membrane (calipers provided for scale).

2. Materials and Methods

2.1. Graphene Nanocomposite Preparation and Characterization

The polymer matrix of the studied nanocomposites consisted of multi-block thermoplastic elastomer consisting of hard segments of poly(ethylene terephthalate) (PET) and soft segments of amorphous fatty acid ester sequences based on di-linoleic acid (DLA). The synthesis of both neat copolymers and nanocomposites was carried out in two stages, based on our previous work [26–28], in a kilo-scale (3.5-dm^3) Fourné Maschinenbau polycondensation reactor. Briefly, the first stage consisted of transesterification of dimethyl terephthalate (DMT) and ethylene glycol (EG), followed by esterification with DLA, and, finally, polycondensation. After the completion of the reaction, the polymer was discharged from the reactor in the form of an extruded filament, cooled with water, and cut into granules suitable for further processing, such as injection or compression molding.

Two different neat co-polymers were prepared with different hard to soft segment ratios, resulting in a more elastic material with 40 wt% of hard segments (PET-DLA 4060) and a stiffer co-polyester with 60 wt% of hard segments (PET-DLA 6040). Additionally, nanocomposites with the same hard to soft segment ratios but with 1 wt% of a commercial graphite nanoplatelet nanofiller Grade-A0-3 by the Graphene Supermarket were prepared via in-situ polymerization. (The producer indicates that the Grade-A0-3 graphene has an average thickness of ~12 nm, representing >30 monolayers; therefore, according to the latest classification [29], the Grade-A0-3 material should be termed as a "graphite nanoplatelet"). The synthesis was carried out in stages, in the same fashion as that of the neat polymer, but the nanofiller was first dispersed in the liquid EG and DLA monomers by a combination of bath and probe sonication and high-speed mixing.

We have already published detailed analyses of these materials [26,27], including Fourier transform infrared spectroscopy (FTIR), proton nuclear magnetic resonance (^{1}H NMR), differential scanning calorimetry (DSC), water contact angle measurement, tensile testing, moisture vapor transmission tests, bacterial adhesion testing, and a cytocompatibility assessment.

For THz-TDS measurements, test samples were prepared using compression molding (Remi-Plast PH10T hot press). For the more elastic PET-DLA 4060 copolymer and nanocomposite specimen, nominally 0.3-mm-thick samples were prepared in the same fashion as the membranes for the prototype heart assist device (Figure 1b): The temperature was 175 °C and the pressure was 6 bar. For the stiffer PET-DLA 6040 copolymer and nanocomposites, the temperature was increased to 230 °C, while pressure remained unchanged, and thicker samples (nominally 0.9 mm thick) were prepared.

After fabrication, our nanocomposite samples were characterized using both a high-resolution TEM (HRTEM) and Raman methods in order to image graphene nanoflakes and estimate their size, dispersion, and a number of graphene layers in a flake. Figure 2 presents representative TEM images obtained using a JEM-2100 JEOL microscope at different magnification levels of the thin, 1-wt% graphene-PET-DLA 4060 nanocomposite sample. Figure 2a is a low-magnification image covering a 10 × 10-μm^2 area showing the density and dispersion of the flakes. In this randomly selected image we can identify 12 nanofiller flakes that are well separated from each other. Besides the one "giant" flake (#1), the other flakes have sizes below 1 μm, so, in rough approximation, they can be treated as independent and non-interacting particles dispersed in a host copolymer matrix. Figure 2b shows a HRTEM image of one of the smallest (7-nm × 130-nm) nanoflakes observed and atomic-scale measurements of interplane distances from two areas that could be well oriented for TEM studies. We can clearly image graphene planes and, in this case, see approximately ten layers. Since the starting nanofiller from Graphene Supermarket consisted of graphite nanoplatelets, HRTEM imaging confirmed that the subsequent process of dispersing this nanofiller in the monomers, followed by in-situ polymerization, had exfoliated the nanoplatelets as expected. The insets showed the individual layers with interlayer distances (calculated based on ten layers) equal to 0.355 nm and 0.362 nm for the top and bottom panels, respectively. In both cases, these values are close to the thickness of a single graphene sheet. We must stress, however, that it is difficult to precisely measure the thickness of the graphene layers in our nanocomposite samples because flakes embedded within the polymer matrix are randomly oriented and their edges are often folded, making proper orientation of the sample during the TEM study difficult. Therefore, although our TEM images and measurements of the thickness of carbon planes in our nanoflakes are consistent with those of graphene monolayers, we cannot completely exclude the probability of having graphite nanoplatelets in our samples.

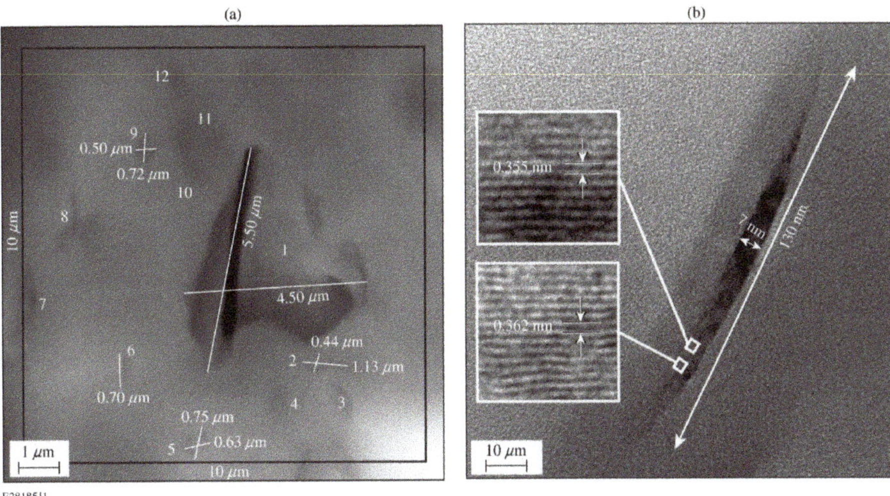

Figure 2. (**a**) Low-resolution (10 × 10-μm^2 area), representative TEM image of 1-wt% graphene PET-DLA 4060 nanocomposite. Twelve (most visible in the view) flakes are identified and the sizes of several flakes are listed. (**b**) HRTEM of the 7-nm × 130-nm nanoflake. Insets show images of graphene planes and interplanar distances from two different sections of the nanoflake (marked as squares).

The above-mentioned ambiguity was resolved by performing scanning fluorescence measurements and Raman spectroscopy. A beam from a helium–neon laser (632.8-nm wavelength) was focused on the sample, and the fluorescence was sent through a 635-nm pass filter (to block the laser line) and was collected by an avalanche photodiode. By scanning the sample using a movable nano-positioning stage, a spatial fluorescence image was created (see inset in Figure 3a). After the scan, the emission was redirected to a spectrometer equipped with a charge-coupled–device (CCD) camera to obtain the Raman spectrum (Figure 3a, main panel). A grating with 600 lines/mm was used to disperse the signal onto the CCD array. The nanoflake under investigation, from which the spectrum was taken, is circled in the inset. The three prominent features visible in the nano-Raman spectrum are the D peak at ~1340 cm^{-1}, the G peak at ~1580 cm^{-1}, and the 2D peak at 2680 cm^{-1}. These features correspond to a few-layer graphene (FLG) sample [29].

We have also performed Raman measurements of an entire 1-wt% graphene-polymer nanocomposite sample, as well as the reference neat PET-DLA 4060 copolymer specimen. Figure 3b shows a spectrum that our graphene nanofiller obtained by numerical subtraction of the neat polymer spectrum from that of the nanocomposite (both shown in the inset in Figure 3b). We note that the positions of the D, G, and 2D peaks very close to that of the nano-Raman study of a single flake. The only difference is the height of the D peak in Figure 3b, which is a defect peak. The discrepancy can be attributed to spectra being collected simultaneously from nanofiller flake edges as well as from within the flake. The edges will increase the observed D band in our spectra. Overall the observed Raman spectra corroborated the fact that, indeed, our nanofiller consisted predominantly of FLG nanoflakes.

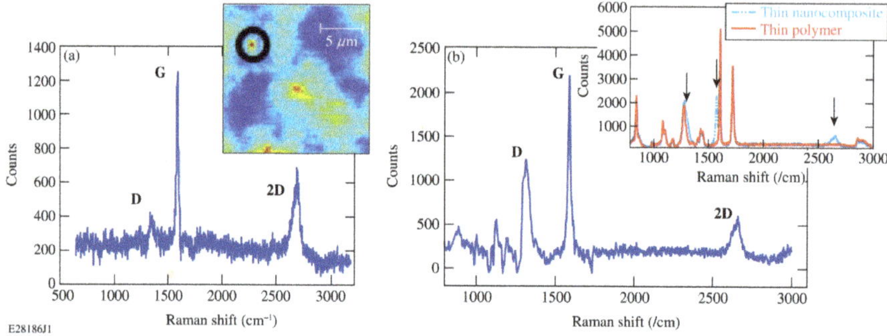

Figure 3. (**a**) Background-subtracted Raman spectrum from a FGL nanoflake embedded in PET-DLA 4060 copolymer matrix. The three prominent features are the D peak at ~1340 cm^{-1}, the G peak at ~1580 cm^{-1}, and the 2D peak at 2680 cm^{-1}. The inset presents a scanning fluorescence image of the region from which the spectrum was taken, with the studied nanoflake circled. For the images, the excitation wavelength was 632.8 nm and the laser power was 600 µW. (**b**) Background-subtracted Raman spectrum of graphene nanoflakes obtained by subtraction of the spectrum of neat PET-DLA 4060 copolymer, used as the graphene nanocomposite matrix. The latter two spectra are both shown in the inset in (**b**), with arrows pointing to the characteristic features of the nanocomposite. The positions of the main peaks in the main panel are the D peak at ~1311 cm^{-1}, the G peak at ~1586 cm^{-1}, and the 2D peak at 2661 cm^{-1}.

2.2. THz Time-Domain Spectroscopy Technique

The THz-TDS system, which is used to measure the THz-range transmission spectra of both the neat copolymers and graphene nanocomposite samples, is schematically presented in Figure 4. THz radiation was generated and detected using a commercial, low-temperature–grown GaAs (LT-GaAs) photo-conductive antenna emitter and detector from TeraVil Ltd., Vilnius, Lithuania [30]. The emitter and detector were excited and probed, respectively, by 100-fs-wide pulses, with 800-nm wavelength and 76-MHz repetition rate, generated by a femtosecond Ti:Sapphire laser. The spectral range of the spectrometer was ~4 THz, with a maximum amplitude of ~0.5 THz. The sample was placed directly between the emitter and detector, as shown in Figure 4, and measurements were taken at room temperature. To reduce the influence of water absorption, the THz emitter, detector, and sample holder were placed inside a Plexiglas® box that was purged with dry nitrogen, ensuring that the humidity during the measurement was below 5%.

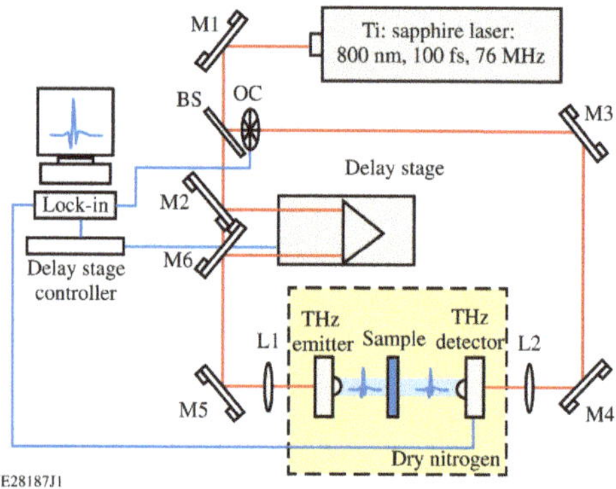

Figure 4. Schematic diagram of a THz-TDS setup. L1, L2: optical lenses; M1–M6: dielectric mirrors; BS: beam splitter, and OC: optical chopper.

For each sample, we first performed a reference run with no sample inside the spectrometer to confirm the performance of the system. Next, we took two measurements; one of a graphene nanocomposite and the other of a corresponding neat co-polymer specimen. To get better results and reduce noise, each set of measurements consisted of at least ten averages. Our measurements were focused on two different sample types: A nominal 0.3-mm-thick, elastic PET-DLA 4060 copolymer and a nominal 0.9-mm-thick, stiff PET-DLA 6040 copolymer. For both sample types, the nanofiller content was the same, i.e., 1 wt%. Since our further analysis crucially depends on the sample thickness, all measurements were repeated several times. At different spots of the test sample, the corresponding local thicknesses were measured, and, subsequently, the results were averaged. Therefore, from now on, we will refer to a nominal 0.3-mm PET-DLA 4060 soft copolymer sample as a "thin sample," and a nominal 0.9-mm PET-DLA 6040 stiff copolymer sample as a "thick sample."

Figure 5a presents the time-domain signals transmitted through dry nitrogen (the reference measurement), the thin and thick neat copolymers, and the corresponding thin and thick, 1-wt% graphene-polymer nanocomposite samples, respectively (see the legend for detailed color/line coding). As expected, the transient corresponding to the thick sample with graphene exhibits the smallest amplitude and the longest arrival time to the detector. The differences in time delays, with respect to the nitrogen signal observed in the signals corresponding to different samples, are caused by their different indexes of refraction, as will be discussed later in connection with Figure 6a. In Figure 5b, we present the corresponding (see the legend) power spectra of our time-domain signals obtained by means of fast Fourier transformation (FFT). We note that, while both copolymers absorb THz radiation, as compared to the dry-nitrogen reference signal, adding graphene flakes to the polymer matrix substantially reduces the bandwidth of the power spectrum (Figure 5b), obviously because of the extra absorption of THz radiation by nanoflakes. The cut-off frequencies for the thin copolymer sample and the corresponding 1-wt% graphene-polymer nanocomposite are ~3.1 THz and 2.9 THz, respectively, while for the thick copolymer and the 1-wt% graphene-polymer nanocomposite they are 2.25 THz, and 1.75 THz, respectively. We observe no resonant absorption features, but rather a broadband THz attenuation.

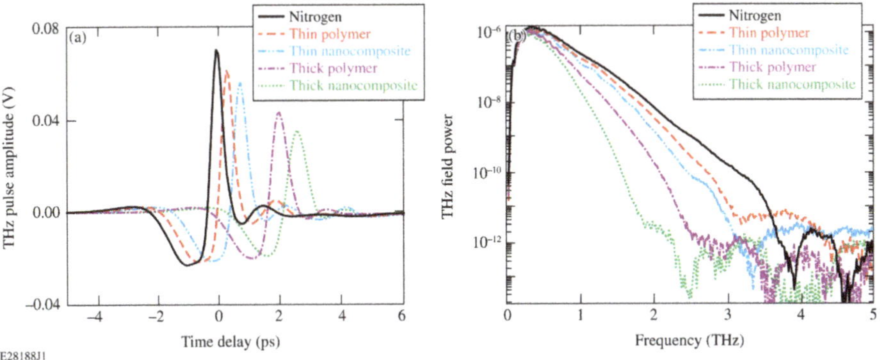

Figure 5. (**a**) Time-resolved transient signals for the tested thin and thick polymers and 1-wt% graphene nanocomposite samples and the empty spectrometer (nitrogen). (**b**) The corresponding power spectra showing absorption of the samples. Note: The thin samples consist of the PET-DLA 4060 copolymer, while the thick ones consist of PET-DLA 6040.

3. Results and Discussion

3.1. Data Analysis

Using Fresnel equations, the THz-TDS approach allows us to find the complex index of refraction $\hat{n}(\omega) = n(\omega) + i\kappa(\omega)$, or the dielectric function $\hat{\varepsilon}(\omega) = \varepsilon_1(\omega) + i\varepsilon_2(\omega)$, of a tested material. The $\hat{\varepsilon}(\omega)$

can be represented as the square of $\hat{n}(\omega)$, i.e., $\hat{\varepsilon}(\omega) = [\hat{n}(\omega)]^2$. When the sample is placed between the THz emitter and detector, the recorded electric field signal $E_{\text{sam}}(t)$ is the THz transient that is transmitted through sample. The reference signal, $E_{\text{ref}}(t)$, is registered when the same measurement is taken, but without the sample. By taking FFT of time-domain transients, we can obtain the frequency-domain spectra of both the sample and reference. When these two spectra are compared, the $\hat{\varepsilon}(\omega)$ and $\hat{n}(\omega)$ dependencies, as well as the absorption coefficient $\alpha(\omega)$, can be calculated.

In agreement with our measurements, we can treat our samples as optically thick materials. In this case, internal reflections do not overlap with the main THz pulse, so, for the sake of simplicity, the data before and after the main pulse can be truncated and only the main THz pulse, analyzed [31]. Consequently, the transmission amplitude ratio $A(\omega)$ and phase difference $\varphi(\omega)$ of sample and reference fields can be calculated as [32],

$$\frac{E_{\text{sam}}(\omega)}{E_{\text{ref}}(\omega)} = \frac{4n(\omega)}{[n(\omega)+1]^2} e^{-\frac{\alpha d}{2} + i\omega\frac{[n(\omega)-1]d}{c}} = A(\omega)e^{i\varphi(\omega)}, \tag{1}$$

where c is the speed of light, d represents the sample thickness, and ω is the angular frequency. From Equation (1) we can derive the absorption coefficient $\alpha(\omega)$ and the real component $n(\omega)$ of $\hat{n}(\omega)$:

$$\alpha(\omega) = -\frac{2}{d}\ln\left\{A(\omega)\frac{[n(\omega)+1]^2}{4n(\omega)}\right\}, \tag{2}$$

$$n(\omega) = 1 + \frac{c}{\omega d}\varphi(\omega). \tag{3}$$

In addition, the extinction coefficient $\kappa(\xi)$ is related to the absorption coefficient as,

$$\kappa(\omega) = \frac{c\alpha(\omega)}{2\omega}. \tag{4}$$

After finding the components of $\hat{n}(\omega)$ and $\hat{\varepsilon}(\omega)$, the complex conductivity $\hat{\sigma}(\omega) = \sigma_1(\omega) + i\sigma_2(\omega)$ of the sample can be subsequently calculated using Equations (5)–(8) [33]:

$$\varepsilon_1(\omega) = n(\omega)^2 - \kappa(\omega)^2, \tag{5}$$

$$\varepsilon_2(\omega) = 2n(\omega)\kappa(\omega), \tag{6}$$

$$\sigma_1(\omega) = 2\varepsilon_0\omega n(\omega)\kappa(\omega), \tag{7}$$

$$\sigma_2(\omega) = \left[\varepsilon_{\text{pl}}(\omega) - n(\omega)^2 + \kappa(\omega)^2\right]\varepsilon_0\omega, \tag{8}$$

where ε_0 is the permittivity of the free space, and $\varepsilon_{\text{pl}}(\omega)$ is the dielectric constant of a neat polymer matrix. We assume that imaginary part ε_{p2} is negligibly small since the neat polymer is an insulating material. $\varepsilon_{\text{pl}}(\omega)$ can be extracted from THz-TDS measurements of a neat polymer in dry nitrogen using the same routine as presented above for the graphene-polymer nanocomposite (using Equations (1)–(5)).

3.2. Index of Refraction and Absorption Coefficient

As mentioned before, we studied two sets of samples: Thin (nominal 0.3 mm) and thick (nominal 0.9 mm); each set contained a different co-polymer as reference and 1-wt% graphene nanocomposite. Since the sample thickness is a very important parameter for calculating $\hat{n}(\omega)$ and $\varepsilon(\omega)$, we repeated our measurements several times at different spots on a given sample and used the averaged value for the sample parameter calculation. Spectral dependencies of $n(\omega)$ and $\alpha(\omega)$ obtained from the THz-TDS data, using Equations (2) and (3), are presented in Figure 6a,b, respectively.

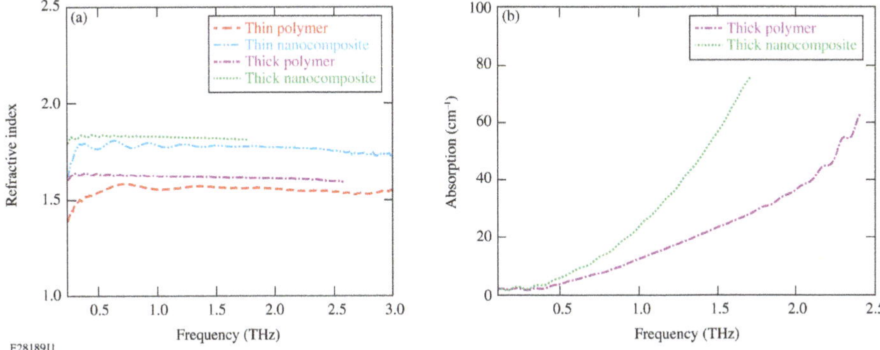

Figure 6. (**a**) Index of refraction and (**b**) absorption spectra for the tested neat copolymer and 1-wt% graphene nanocomposite samples. Note: The thin sample consisted of PET-DLA 4060 copolymer, while the thick one consisted of PET-DLA 6040. The data points were trimmed according to sample bandwidth (see Figure 5b).

One should ignore the low-frequency oscillations in the extracted $n(\omega)$ and $\alpha(\omega)$ spectra (Figure 6), that are the most pronounced for the thin samples because these oscillations can be attributed to the approximation error of our simplified model. The methodology behind Equation (1) requires that the time-domain signal be truncated before any secondary peaks (because of internal reflections). This truncation, as well as minor deviation or inaccuracy in the sample thickness determination, introduces an incomplete removal of the Fabry–Pérot reflections of the THz signal, resulting in observed oscillations. Note that the oscillations for the thick samples are much smaller and have a smaller period, which is consistent with this approximation error. These errors, while relatively small in the $n(\omega)$ spectra, were very substantial in the calculated $\alpha(\omega)$ spectrum for the thin samples and, eventually, prevented us from obtaining a consistent plot in the latter case.

The $n(\omega)$ spectra of both neat copolymers, as seen in Figure 6a, are almost flat across the entire measured range, and the values at 1 THz for a thin and a thick polymer are ~1.57 and ~1.62, respectively, which are close to the values observed earlier for PET polymer [34,35]. The different $n(\omega)$ values for the thin and thick samples are caused by the different ratio between the hard and soft segments in the PET-DLA 4060 and PET-DLA 6040 copolymers. The PET-DLA 6040 co-polymer has a higher percentage of PET hard segments and therefore a higher crystallinity, resulting in the increased n value. When 1 wt% of graphene nanofiller was added into each copolymer, the $n(\omega)$ dependences of the resultant nanocomposites remained flat in the measurement range, but the values, again evaluated at 1 THz, increased to ~1.78 and ~1.83 for the thin and thick samples, respectively. Clearly, adding 1 wt% of graphene flakes to a copolymer matrix measurably increased (~10%) the n of the resultant nanocomposite. Similar results were obtained by other groups [21,22].

As explained above, Figure 6b shows $\alpha(\omega)$ spectra for only the thick samples, i.e., the neat PET-DLA 6040 co-polymer and the corresponding graphene nanocomposite. The addition of 1-wt% graphene into the copolymer matrix increases the $\alpha(\omega)$, especially at the high-frequency end of the spectrum. The latter result is in good agreement with previously published data [22].

3.3. Dielectric Constant

Knowing the $n(\omega)$ dependence, it is easy to calculate the spectra for the relative dielectric constant $\varepsilon_1(\omega)$ for both the neat co-polymer and the graphene nanocomposite, for both the thin and thick samples, as shown in Figure 7a,b, respectively. The resulting $\varepsilon_1(\omega)$ spectra are flat, as we observed earlier for the $n(\omega)$ dependences presented in Figure 6a. The insertion of graphene nanoflakes increases the dielectric constant of the corresponding nanocomposite sample. Despite a very large frequency difference, our numerical values of $\varepsilon_1(\omega)$ are close to those reported by Marra et al. [21], measured at radio frequencies from 8 to 12 GHz.

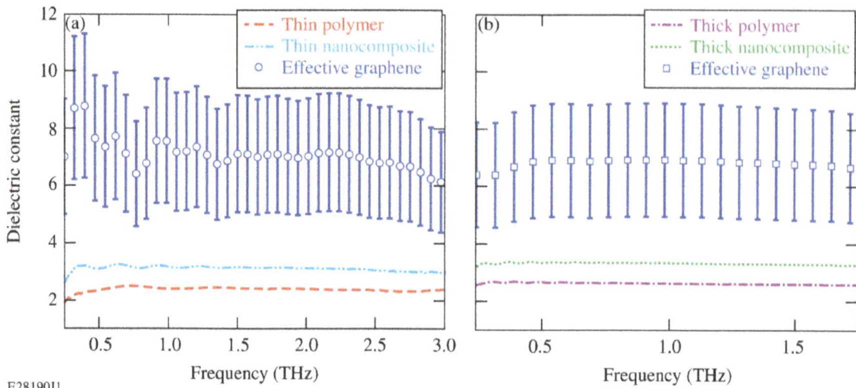

Figure 7. Real part of the complex dielectric function spectra (solid lines) for (**a**) thin and (**b**) thick polymer and 1-wt% graphene nanocomposite samples. Note: The thin samples consist of PET-DLA 4060 copolymer, while the thick one was PET-DLA 6040. The open circle and open square symbols present the effective graphene flake dielectric function based on the Clausius–Mossotti relation. The error bars are ~30%.

The $\varepsilon_1(\omega)$ spectra presented in Figure 7 were subsequently used to estimate the dielectric constant of only graphene flakes embedded in the polymer. Our approach is based on the classical Clausius–Mossotti relation that describes the dielectric constant of a composite sample ε_c [36] as

$$\varepsilon_c(\omega) = \frac{3\varepsilon_{\mathrm{pl}}(\omega) + 2N\alpha}{3\varepsilon_{\mathrm{pl}}(\omega) - N\alpha}\varepsilon_{\mathrm{pl}}(\omega), \tag{9}$$

where N represents the density of the graphene nanoflakes and α in this case, is a so-called "atomic polarizability". We note here that the effective-medium theory of Bruggeman, most commonly used for the composite material analysis, are not applicable here since it requires that volume fraction of a filler material be at least 10% [37]. This condition is, obviously, not fulfilled in our case because we have only 1 wt% of graphene nanoflakes in the copolymer matrix. On the other hand, based on the TEM images (see Figure 2a), the graphene nanoflakes in our samples are well separated from each other. Consequently, we can assume that they act as independent and noninteracting particles in a host material and use the Clausius–Mossotti relation that is the exact solution of a dielectric constant of a medium having classically interacting dipoles embedded in an insulating matrix. Therefore, we simply treat each flake as a dipole and the dielectric constant of graphene nanoflakes $\varepsilon_{gr}(\omega)$ should be proportional to $N\alpha$ and given as

$$\varepsilon_{gr}(\omega) = N\alpha * \frac{1}{\delta}, \tag{10}$$

where δ is the volume ratio of graphene nanoflakes to the entire sample volume. Of course, the 1-wt% ratio does not reflect the volume ratio, so we used the TEM images of our samples (see Figure 2a) to estimate the volume percentage of the randomly distributed and orientated graphene nanoflakes inside our nanocomposite. The calculation results of $\varepsilon_{gr}(\omega)$ are shown in Figure 7 as symbols. The error bars represent an uncertainty of ~30% in our evaluation of δ. Comparing the results presented in Figure 7a,b, we see that ignoring the low-frequency section of $\varepsilon_{gr}(\omega)$ for the thin sample (mentioned in earlier Fabry–Pérot reflections), the $\varepsilon_{gr}(\omega)$ values for both graphene nanocomposite samples are very close to each other and are ~7 ± 2.

3.4. Complex Conductivity

The complex conductivity calculations are based on Equations (7) and (8). In order to calculate the complex $\hat{\sigma}(\omega)$ of our graphene nanocomposite samples, we need to know the complex dependences of $\hat{\varepsilon}(\omega)$ for not only the nanocomposite, but also for the neat polymer matrix. Therefore, we additionally calculated the $\hat{\varepsilon}(\omega)$ spectrum for each neat copolymer sample, using Equations (1)–(6), with the reference corresponding to the THz transient spectrum measured in dry nitrogen. The resulting real and imaginary parts of the graphene nanocomposite complex conductivity are plotted as symbols for both the thin (open circles) and thick (open triangles) in Figure 8. To calculate error bars, we analyzed the data collected in different sets of measurements performed at different spots on our samples.

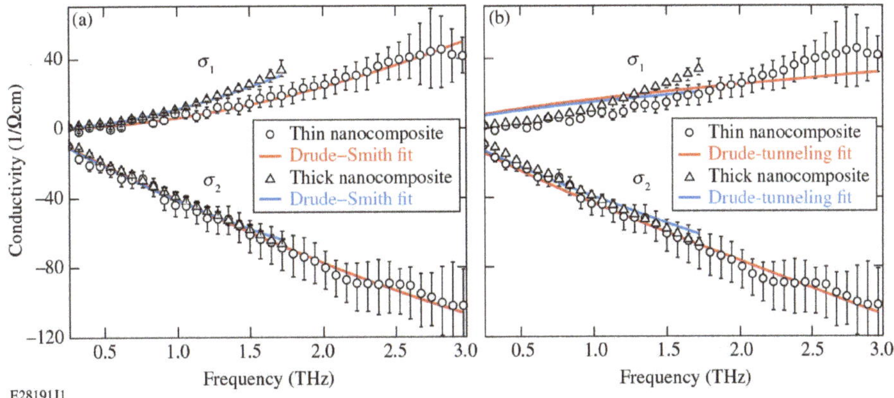

Figure 8. Experimental complex conductivity (σ_1 and σ_2) spectra for both thin (open circles) and thick (open triangles), 1-wt% graphene nanocomposite samples. Solid lines correspond to the fits based (**a**) the Drude–Smith and (**b**) Drude-tunneling models. Experimental data points for the thick sample were trimmed according to sample bandwidth (see Figure 5b).

Next, our $\hat{\sigma}(\omega)$ spectra were fitted using the Drude–Smith model [38] (solid lines in Figure 8a), and what we call the Drude-tunneling model presented in [39] (solid lines in Figure 8b). The Drude–Smith model is given by

$$\sigma(\omega) = \frac{\varepsilon_0 \omega_p^2 \tau}{(1 - i\omega\tau)} \left[1 + \sum_{m=1}^{\infty} \frac{c_m}{(1 - i\omega\tau)} \right], \tag{11}$$

where ω_p is the plasma frequency, τ represents the carrier collision time, and the c_m coefficient is the so-called scattering parameter. A first-order approximation is conventionally used, i.e., the terms $m > 1$ are truncated. When $c_m = 0$, Equation (11) simplifies to the classical Drude model (the term in front of the square bracket), while $c_1 = -1$ means that carriers are completely backscattered. This extension of the classic Drude model was introduced by Smith [38] and is usually used to describe complex conductivity spectra in the THz range of materials with nanofillers [40], such as nanoparticles [41,42], carbon nanotubes [43], and even monolayer graphene [43], when it is difficult to explain results using the simple, homogeneous-medium Drude model.

The model presented in [39] is based on a concept that in non-homogeneous, granular materials carrier transport consists of both, intragrain scattering and intergrain tunneling, and the corresponding $\sigma_{eff}(\omega)$ should be described, in general, as a superposition of both processes and given by as

$$\frac{1}{\sigma_{eff}(\omega)} = \frac{f}{\sigma_f(\omega)} + \frac{1-f}{\sigma_t(\omega)}, \tag{12}$$

where $\sigma_f(\omega)$ is the free carrier conductivity given by the Drude law ($m = 0$ in Equation (11)), $\sigma_t(\omega)$ is the tunneling conductivity given in [39], and f is the spectral weight of intragrain transport. The latter explains why the authors in [39] call this approach the free (intra-grain) and tunneling (inter-grain; grain boundary) carrier transport model.

Analyzing Figure 8a, we observe that we obtained an excellent fit using, as it is a standard procedure, only the first term ($m = 1$) in the Smith expansion (Equation (11)) and the fitting parameters were $c_1 = -0.997$, $\omega_p = 72.4 \times 10^{12}$ 1/s, and $\tau = 12$ fs for the thin nanocomposite sample, and $c_1 = -0.998$, $\omega_p = 41.9 \times 10^{12}$ 1/s, and $\tau = 21$ fs for the thick one (in both cases, the $\omega_p \tau$ product is approximately the same and equal to 0.87). We observe that in both cases, the c_1 values are very close to -1, indicating a nearly complete backscattering mechanism of carriers inside our graphene-polymer samples.

In the case of Figure 8b, the fit based on the Drude-tunneling model (Equation (12)) is not as good, especially in the case of σ_1, and we need to stress that it was obtained for $f = 0.98$, i.e., the tunneling process is almost negligible in our case. The latter, combined with the fact that for both the thin and thick nanocomposite samples, σ_2 is negative in the entire frequency range, points to that in our nanocomposites (based on either PET-DLA 4060 or PET-DLA 6040 copolymers) the graphene nanofiller is well-distributed, and consists of individual conducting nanoflakes that are well isolated from each other inside a polymer matrix.

Using the Drude–Smith best-fit ω_p parameters, and by experimentally-determining the graphene layer-to-layer separation distance (see Figure 2b), we estimated the free-carrier concentration in our nanoflakes to be equal to ~1.1 $\times$ 10^9 cm^{-2} and ~3.2 $\times$ 10^9 cm^{-2} for the thick, and thin sample, respectively. The ultrashort carrier collision time τ obtained from the Drude–Smith model indicates, in turn, that our graphene nanoflakes are, on average, much larger than the carrier mean-free path within the flake, estimated to be on the order of 10 nm. The latter confirms the applicability of the Drude-Smith approach to model the carrier transport in our graphene-polymer nanocomposites. Finally, we point to very similar THz conductivity studies, very recently reported by Skalsky et al. [22]. The authors used samples consisting of FLG flakes embedded in polytetrafluoroethylene (PTFE) pellets, and measured them using the THZ-TDS method. They fitted only the σ_1 spectrum, but also obtained c_1 very close to -1 and very similar free-carrier concentration, as well as came to very similar conclusions as ours.

4. Conclusions

We have successfully used THz time-domain spectroscopy to characterize multi-block, co-polyester, graphene nanocomposite materials, synthesized using kilo-scale (suitable for Research and Development, and prototyping) in-situ polymerization and processed using industrially/economically viable compression molding. We present an experimental procedure and step-by-step data workup method used to characterize the FLG nanofiller within an insulating matrix material for the case where the filler is at relatively low density (1 wt%). We have shown that one can extract the refractive index, the dielectric constant, and the complex conductivity of the graphene-polymer nanocomposite material, as well as the effective dielectric constant of graphene nanoflakes within the matrix. In the THz-TDS method, the dielectric properties represent a sample average of all the materials in an almost cm^2 cross-section beam path; consequently, these data contain "global" information regarding the dispersion of the filler (in our case FLG) and its final retained "quality." In our study, the Drude–Smith backscatter parameter, which is close to -1, and the negative imaginary part of the complex conductivity both indicate that our nanofiller flakes are fully isolated in the polymer matrix. The quality of the dispersion is implied by the high value of the conductivity and moderate effective dielectric constant that is retained by the graphene nanoflakes. High THz conductivity typically corresponds to high carrier mobility within individual graphene flakes, which in turn suggests that flakes are structurally intact, without a significant amount of scattering defects introduced. The results of our THz study are consistent with the HRTEM image analysis, and indicate that THz-TDS characterization is uniquely suited to probe the dispersion and in-situ electronic quality of nanofillers such as graphene or, e.g., carbon-nanotubes in various, not only polymer, composite materials. In the latter context, however, it is important to stress that tested elastic PET-DLA 4060 copolymer nanocomposites are analogous to experimental hemispherical pneumatic membranes of a prototype extra-corporeal heart assist device (Figure 1). Consequently, this work, as well as other similar THz studies, serves as an important proof of concept of THz spectroscopy of nanocomposites in the product development chain. Non-destructive measurements could be carried out at different stages of the process: Aafter polymer nanocomposite synthesis, following processing into a given prototype component, and even after "field" testing (either mechanical or functional).

Author Contributions: Conceptualization, M.E.F. and R.S.; methodology, A.K., Z.S., P.S., and C.P.R.; investigation, A.K., G.C., K.M.G., Z.S., and A.Ł.; formal analysis, A.K., G.C. P.S., and C.P.R.; validation, A.N.V., M.E.F., A.C., and R.S.; visualization, A.K., G.C., K.M.G., and A.Ł.; resources, A.N.V., M.E.F., A.C., and R.S.; writing—original draft preparation, A.K., G.C., P.S., C.P.R., and R.S.; writing—review and editing, A.K., A.N.V., P.S., M.E.F., A.Ł., C.P.R., and R.S.; supervision, R.S.; project administration, A.N.V., M.E.F., A.C., and R.S.; funding acquisition, A.N.V., M.E.F., A.C., and R.S.

Funding: This research was funded in part by the PumpPrimerII Program at the University of Rochester. A.N.V. acknowledges support from the Air Force Office of Scientific Research (FA9550-16-1-0020). M.E.F. acknowledges support from the Polish National Centre for Research and Development (PBS1/A5/2/2012).

Acknowledgments: P. Sobolewski thanks W. Ignaczak from the West Pomeranian University of Technology for their helpful discussions.

References

1. Paul, D.R.; Robeson, L.M. Polymer nanotechnology: Nanocomposites. *Polymer* **2008**, *49*, 3187–3204. [CrossRef]
2. Zhu, Y.; Murali, S.; Cai, W.; Li, X.; Suk, J.W.; Potts, J.R.; Ruoff, R.S. Graphene and graphene oxide: Synthesis, properties, and applications. *Adv. Mater.* **2010**, *22*, 3906–3924. [CrossRef] [PubMed]
3. Potts, J.R.; Dreyer, D.R.; Bielawski, C.W.; Ruoff, R.S. Graphene-based polymer nanocomposites. *Polymer* **2011**, *52*, 5–25. [CrossRef]
4. Hu, K.; Kulkarni, D.D.; Choi, I.; Tsukruk, V.V. Graphene-polymer nanocomposites for structural and functional applications. *Prog. Polym. Sci.* **2014**, *39*, 1934–1972. [CrossRef]
5. Silva, M.; Alves, N.M.; Paiva, M.C. Graphene-polymer nanocomposites for biomedical applications. *Polym. Adv. Technol.* **2018**, *29*, 687–700. [CrossRef]
6. Sobolewski, P.; Piwowarczyk, M.; Fray, M.E. Polymer-graphene nanocomposite materials for electrochemical biosensing. *Macromol. Biosci.* **2016**, *16*, 944–957. [CrossRef] [PubMed]
7. Cui, Y.; Kundalwal, S.I.; Kumar, S. Gas barrier performance of graphene-polymer nanocomposites. *Carbon* **2016**, *98*, 313–333. [CrossRef]
8. Kumar, S.K.; Benicewicz, B.C.; Vaia, R.A.; Winey, K.I. 50th anniversary perspective: Are polymer nanocomposites practical for applications? *Macromolecules* **2017**, *50*, 714–731. [CrossRef]
9. Khare, H.S.; Burris, D.L. A quantitative method for measuring nanocomposite dispersion. *Polymer* **2010**, *51*, 719–729. [CrossRef]
10. Schaefer, D.W.; Justice, R.S. How nano are nanocomposites? *Macromolecules* **2007**, *40*, 8501–8517. [CrossRef]
11. Miltner, H.E.; Watzeels, N.; Goffin, A.-L.; Duquesne, E.; Benali, S.; Dubois, P.F.; Rahier, H.; Van Mele, B. Quantifying the degree of nanofiller dispersion by advanced thermal analysis: Application to polyester nanocomposites prepared by various elaboration methods. *J. Mater. Chem.* **2010**, *20*, 9531–9542. [CrossRef]
12. Espinoza-González, C.; Ávila-Orta, C.; Martínez-Colunga, G.; Lionetto, F.; Maffezzoli, A. A measure of CNTs dispersion in polymers with branched molecular architectures by UDMA. *IEEE Trans. Nanotechnol.* **2016**, *15*, 731–737. [CrossRef]
13. Zhang, Q.; Fang, F.F.; Zhao, X.P.; Li, Y.; Zhu, M.; Chen, D. Use of dynamic rheological behavior to estimate the dispersion of carbon nanotubes in carbon nanotube/polymer composites. *J. Phys. Chem. B* **2008**, *112*, 12606–12611. [CrossRef] [PubMed]

14. Müller, K.-H.; Bugnicourt, E.; Latorre, M.; Jorda, M.; Echegoyen Sanz, Y.; Lagaron, J.; Miesbauer, O.; Bianchin, A.; Hankin, S.; Bölz, U.; et al. Review on the processing and properties of polymer nanocomposites and nanocoatings and their applications in the packaging, automotive and solar energy fields. *Nanomaterials* **2017**, *7*, 74. [CrossRef] [PubMed]

15. Naftaly, M.; Miles, R.E. Terahertz time-domain spectroscopy for material characterization. *Proc. IEEE* **2007**, *95*, 1658–1665. [CrossRef]

16. Wietzke, S.; Jansen, C.; Reuter, M.; Jung, T.; Kraft, D.; Chatterjee, S.; Fischer, B.M.; Koch, M. Terahertz spectroscopy on polymers: A review of morphological studies. *J. Mol. Struct.* **2011**, *1006*, 41–51. [CrossRef]

17. Amenabar, I.; Lopez, F.; Mendikute, A. In introductory review to THz non-destructive testing of composite mater. *J. Infrared Millim. Terahertz Waves* **2013**, *34*, 152–169. [CrossRef]

18. Stoik, C.; Bohn, M.; Blackshire, J. Nondestructive evaluation of aircraft composites using reflective terahertz time domain spectroscopy. *NDT&E Int.* **2010**, *43*, 106–115.

19. Ospald, F.; Zouaghi, W.; Beigang, R.; Matheis, C.; Jonuscheit, J.; Recur, B.; Guillet, J.-P.; Mounaix, P.; Vleugels, W.; Bosom, P.N.; et al. Aeronautics composite material inspection with a terahertz time-domain spectroscopy system. *Opt. Eng.* **2013**, *53*, 031208. [CrossRef]

20. Casini, R.; Papari, G.; Andreone, A.; Marrazzo, D.; Patti, A.; Russo, P. Dispersion of carbon nanotubes in melt compounded polypropylene based composites investigated by THz spectroscopy. *Opt. Express* **2015**, *23*, 18181–18192. [CrossRef]

21. Marra, F.; D'Aloia, A.; Tamburrano, A.; Ochando, I.; De Bellis, G.; Ellis, G.; Sarto, M. Electromagnetic and dynamic mechanical properties of epoxy and vinylester-based composites filled with graphene nanoplatelets. *Polymers* **2016**, *8*, 272. [CrossRef]

22. Skalsky, S.; Molloy, J.; Naftaly, M.; Sainsbury, T.; Paton, K.R. Terahertz time-domain spectroscopy as a novel metrology tool for liquid-phase exfoliated few-layer graphene. *Nanotechnology* **2018**, *30*, 025709. [CrossRef] [PubMed]

23. Chamorro-Posada, P.; Vázquez-Cabo, J.; Rubiños-López, O.; Martín-Gil, J.; Hernández-Navarro, S.; Martín-Ramos, P.; Sánchez-Arévalo, F.M.; Tamashausky, A.V.; Merino-Sánchez, C.; Dante, R.C. THz TDS study of several sp^2 carbon materials: Graphite, needle coke and graphene oxides. *Carbon* **2018**, *98*, 484–490. [CrossRef]

24. Whelan, P.R.; Huang, D.; Mackenzie, D.; Messina, S.A.; Li, Z.; Li, X.; Li, Y.; Booth, T.J.; Jepsen, P.U.; Shi, H.; et al. Conductivity mapping of graphene on polymeric films by terahertz time-domain spectroscopy. *Opt. Express* **2018**, *26*, 17,748–17,754. [CrossRef] [PubMed]

25. El Fray, M.; Czugala, M. Polish artificial heart program. *WIREs Nanomed. Nanobiotechnol.* **2012**, *4*, 322–328. [CrossRef] [PubMed]

26. Staniszewski, Z.; El Fray, M. Influence of thermally exfoliated graphite on physicochemical, thermal and mechanical properties of copolyester nanocomposites. *Polimery* **2016**, *61*, 482–489. [CrossRef]

27. Staniszewski, Z.; Sobolewski, P.; Piegat, A.; El Fray, M. The effects of nano-sized carbon fillers on the physico-chemical, mechanical, and biological properties of polyester nanocomposites. *Eur. Polym. J.* **2018**, *107*, 189–201. [CrossRef]

28. Piegat, A.; El Fray, M. Polyethylene terephthalate modification with the monomer from renewable resources. *Polimery* **2007**, *52*, 885–888. [CrossRef]

29. Kauling, A.P.; Seefeldt, A.T.; Pisoni, D.P.; Pradeep, R.C.; Bentini, R.; Oliveira, R.V.B.; Novoselov, K.S.; Castro Neto, A.H. The worldwide graphene flake production. *Adv. Mater.* **2018**, *30*, 1803784. [CrossRef]

30. Geižutis, A.; Krotkus, A.; Bertulis, K.; Molis, G.; Adomavičius, R.; Urbanowicz, A.; Balakauskas, S.; Valaika, S. Terahertz radiation emitters and detectors. *Opt. Mater.* **2008**, *30*, 786–788. [CrossRef]

31. Duvillaret, L.; Garet, F.; Coutaz, J.-L. Highly precise determination of optical constants and sample thickness in terahertz time-domain spectroscopy. *Appl. Opt.* **1999**, *38*, 409–415. [CrossRef] [PubMed]

32. Fischer, B.M.; Helm, H.; Jepsen, P.U. Chemical recognition with broadband THz spectroscopy. *Proc. IEEE* **2007**, *95*, 1592–1604. [CrossRef]

33. Zou, X.; Shang, J.; Leaw, J.; Luo, Z.; Luo, L.; La-o-vorakiat, C.; Cheng, L.; Cheong, S.A.; Su, H.; Zhu, J.-X.; et al. Terahertz conductivity of twisted bilayer graphene. *Phys. Rev. Lett.* **2013**, *110*, 067401. [CrossRef] [PubMed]

34. Jin, Y.S.; Kim, G.J.; Jeon, S.G. Terahertz dielectric properties of polymers. *J. Korean Phys. Soc.* **2006**, *49*, 513–517.

35. Fedulova, E.V.; Nazarov, M.M.; Angeluts, A.A.; Kitai, M.S.; Sokolov, V.I.; Shkurinov, A.P. Studying of dielectric properties of polymers in the terahertz frequency range. *Proc. SPIE* **2012**, *8337*, 83370I.

36. Talebian, E.; Talebian, M. A general review on the derivation of Clausius–Mossotti relation. *Optik* **2013**, *124*, 2324–2326. [CrossRef]

37. Scheller, M.; Jansen, C.; Koch, M. Applications of effective medium theories in the terahertz regime. In *Recent Optical and Photonic Technologies*; Kim, K.Y., Ed.; IntechOpen: London, UK, 2010; pp. 231–250.

38. Smith, N.V. Classical generalization of the Drude formula for the optical conductivity. *Phys. Rev. B* **2001**, *64*, 155106. [CrossRef]

39. Shimakawa, K.; Kasap, S. Dynamics of carrier transport in nanoscale materials: Origin of non-Drude behavior in the terahertz frequency range. *Appl. Sci.* **2016**, *6*, 50. [CrossRef]

40. Lloyd-Hughes, J.; Jeon, T.-I. A review of the terahertz conductivity of bulk and nano-materials. *J. Infrared Millim. Terahertz Waves* **2012**, *33*, 871–925. [CrossRef]

41. Richter, C.; Schmuttenmaer, C.A. Exciton-like trap states limit electron mobility in TiO_2 nanotubes. *Nat. Nanotechnol.* **2010**, *5*, 769–772. [CrossRef]

42. Chen, G.; Shrestha, R.; Amori, A.; Staniszewski, Z.; Jukna, A.; Korliov, A.; Richter, C.; Fray, M.E.; Krauss, T.; Sobolewski, R. Terahertz time-domain spectroscopy characterization of carbon nanostructures embedded in polymer. *J. Phys. Conf. Ser.* **2017**, *906*, 012002. [CrossRef]

43. Dadrasnia, E.; Lamela, H.; Kuppam, M.B.; Garet, F.; Coutza, J.-L. Determination of the DC electrical conductivity of multiwalled carbon nanotube films and graphene layers from noncontact time-domain terahertz measurements. *Adv. Condens. Matter Phys.* **2014**, *2014*, 370619. [CrossRef]

Article

Influence of Thickness and Lateral Size of Graphene Nanoplatelets on Water Uptake in Epoxy/Graphene Nanocomposites

Silvia G. Prolongo *, Alberto Jiménez-Suárez, Rocío Moriche and Alejandro Ureña

Materials Science and Engineering Area, Universidad Rey Juan Carlos, Calle Tulipan, s/n,
28933 Móstoles, Spain; alberto.jimenez.suarez@urjc.es (A.J.-S.); rocio.moriche@urjc.es (R.M.);
alejandro.urena@urjc.es (A.U.)
* Correspondence: silvia.gonzalez@urjc.es; Tel.: +34-914-888-292

Received: 25 July 2018; Accepted: 30 August 2018; Published: 4 September 2018

Abstract: In this study, the hydrothermal resistance of an epoxy resin (aircraft quality) reinforced with graphene is analyzed. Different geometries and aspect ratios (thickness and lateral dimensions) of graphene nanoplatelets were studied. The addition of these graphene nanoplatelets induces important advantages, such as an increase of the glass transition temperature and stiffness and an enhancement of barrier properties of the epoxy matrix, in spite of the excellent behavior of pristine resin. The effectiveness of graphene nanoplatelets increases with their specific surface area while their dispersion degree is suitable. Thinner nanoplatelets tend to wrinkle, decreasing their efficiency as nanofillers. Graphene used as reinforcement not only reduces the absorbed moisture content but also decreases its effect on the thermal and mechanical properties related to the matrix.

Keywords: graphene nanoplatelet; epoxy composite; water absorption

---

1. Introduction

Carbon-based nanofillers, such as carbon nanotubes (CNTs), graphene nanosheets (G) or/and graphene nanoplatelets (GNPs) are being widely investigated as reinforcements of epoxy resins [1–4]. Their excellent electrical, thermal and mechanical properties induce interesting expectations of improving the behavior of these thermosetting resins. Among other advantages, the high aspect ratio of these nanofillers together with their hydrophobic character allow them to act as efficient barriers against water transport through the resin. Some authors [5–7] have reported that the reduction of water permeability is associated with both the reduction of the free volume and the restriction of the molecular dynamics of the polymer chain surrounding the nanofiller. Several works [5–10] have been published related to the reduction of the maximum water absorbed caused by the addition of CNTs or GNPs. This effect can be enhanced by the increase of the nanofiller content added into the matrix and its dispersion degree.

On the other hand, it is well known that the absorption of water into thermosetting resins causes a plasticization effect [11,12]. This is due to the fact that water increases the chain segments mobility, which implies a decrease of the glass transition temperature and even the reduction of modulus and mechanical strength. It is expected that the nanofillers addition acts in two different ways: Reducing the rate and/or the maximum content of water uptake and improving the mechanical properties of non-aged composites.

Recently, O. Starkova et al. [5,6] studied the moisture uptake and its effect on the thermo-mechanical properties in composites reinforced with CNTs and reduced graphene oxide. They concluded that the CNTs addition decreases the diffusivity rate but the maximum water absorption remains unchanged. However, the addition of graphene oxide nanoparticles reduces the water absorption capability of the epoxy resin. This means that the geometry of graphitic nanofiller is significant as the GNPs seems to be more efficient. CNTs usually have a higher specific surface area than GNPs. Theoretically, the high specific area of the nanofiller should increase the barrier properties but the higher aspect of CNTs decreases this phenomenon. In addition, graphene nanoparticles tend to a self-orientation, showing a preferential orientation through the in-plain direction, which enhances barrier properties.

In this paper, the influence of GNPs geometry on the moisture uptake of epoxy/GNPs nanocomposites is analyzed. In addition, we analyze the plasticization effect caused by water in the neat epoxy resin and nanocomposites in order to evaluate the influence of the GNPs addition.

2. Materials and Methods

The epoxy matrix was based on a monomer denominated Araldite LY556, cured with an aromatic amine Araldite XB3473. Both components were provided by Antala Group, Cataluña, Spain. The curing was carried out at 140 °C for 8 h, reaching a glass transition temperature (T_g) of 162–165 °C, measured by dynamic mechanical thermal analysis (DMTA).

In this work, different graphene nanoplatelets have been used as nanofiller, the commercial denominations are AO1, AO2, AO3 and AO4, named in this work as GNP1, GNP2, GNP3 and GNP4, respectively. All of them were supplied by Graphene Laboratories Inc, Calverton, NY, USA with a purity close to 99%. Their main geometric characteristics are collected in Table 1. Dimensions were supplied by the manufacturer and corroborated by transmission electron microscopy (TEM).

Table 1. Main geometric characterizes of graphene nanoplatelets.

Graphene Type	Purity (%)	Specific Surface Area (m/g²)	Average Flake Thickness (nm)	Average Particle Lateral Dimension (μm)
AO1 (GNP1)	98.0	510	1.6 (less than 3 layers)	~10
AO2 (GNP2)	99.9	100	8 (20–30 layers)	0.55 (0.15–3)
AO3 (GNP3)	99.2	80	12 (30–50 layers)	4.5 (1.5–10)
AO4 (GNP4)	98.5	<15	60 (~180 layers)	3–7

The manufacture of GNP/epoxy composites has been optimized in previous works [13,14]. The procedure consists in GNPs dispersion, in all the proposed cases in a weight content of 0.5%, into the neat epoxy monomer using a high-speed mixer (Dispermat; Lumaquin, S.A, Barcelona, Spain) at 6000 rpm for 15 min to obtain the optimum doughnut effect. In order to remove the trapped air, once dispersion was completed, the mixture was degassed under vacuum at 80 °C for 15 min. Then, the stoichiometric ratio of the amine curing agent was added at 80 °C and the curing treatment was applied in an oven.

The morphological study of pristine graphene nanoparticles and epoxy composites reinforced with different GNPs was carried out by Transmission Electron Microscopy (TEM, Phillips Tecnai 20 of 200 kV, FEI Company, Hillsboro, OR, USA) and Field-Emission Gun Scanning Electron Microscopy (FEG-SEM, Nova NanoSEM FEI 230, FEI Company). For electron microscopy (TEM and FEG-SEM), epoxy samples were cut by cryomicrotomy. In addition, the film was coated with a thin layer (5–10 nm) of Au for FEG-SEM observation. The experimental conditions of the sputtering were 30 mA for 120 s (Bal-tec, SCD-005 sputter).

Bar-shape samples were cut with different dimensions: $35 \times 12 \times 1.5$ mm³ for DMTA (dynamic thermomechanical analysis) measurements and $60 \times 12.5 \times 2$ mm³ for flexural test. In order to carry out the study, samples were introduced in a hydrothermal chamber with a relative humidity of 85% and a temperature of 40 °C. The moisture absorption was determined by gravimetric method.

The samples were periodically removed and weighted with an accuracy of 0.01 mg. At selected times, some samples were tested by DMTA and flexural test.

The mechanical characterization of composites was carried out by flexural test (ZwickLine Z2.5, Zwick-Roell, Ulm, Germany), following the ASTM D-790 (procedure A and B). DMTA was performed following the standard D5418-01, using the single cantilever bending mode in a DMTA Q800 V7.1 from TA Instruments, New Castle, DE, USA. All the experiments were carried out at 1 Hz frequency, by bending deformation, and scanning from 20 to 250 °C with a heating rate of 2 °C/min.

3. Results

3.1. Morphology

Figure 1 shows high resolution micrographs of the different types of the graphene nanoplatelets used in this work. It can be clearly seen that the thickest platelets are GNP4 and the thinnest ones are GNP1, according to the datasheets of the manufacturer (Table 1). Despite not showing a cross-section of GNP1, the fact that the thickness of GNP1 is the lowest of the four types is attributed to the lack of different contrast between the nanoplatelet and the background shown in Figure 1a, elucidating the low number of layers. The average lateral dimension is difficult to determine, but it is possible to observe that GNP2 nanoplatelets are smaller, while the largest ones are GNP4. Figure 2 shows the schematic morphology of each type of graphene nanoplatelet used in the study.

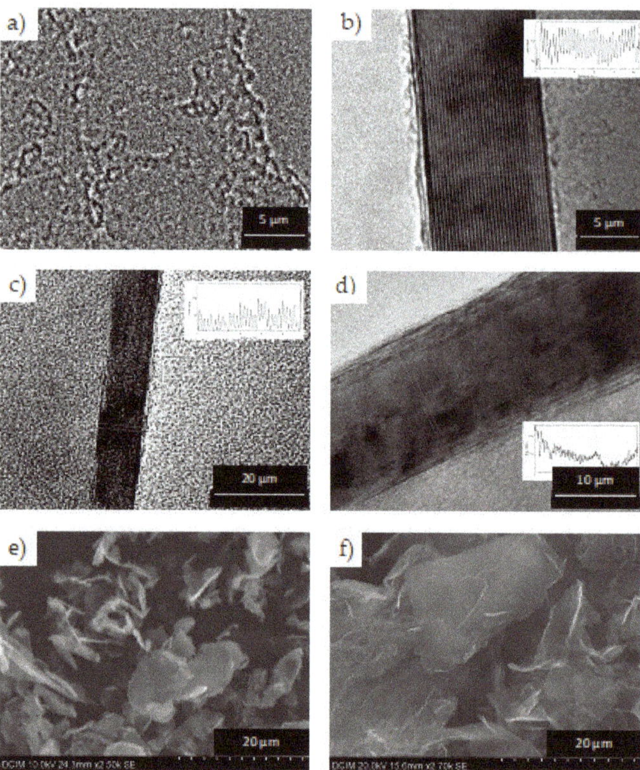

Figure 1. TEM micrographs of (**a**) GNP1; (**b**) GNP2; (**c**) GNP3; (**d**) GNP4 and SEM micrographs of (**e**) GNP2 and (**f**) GNP4.

The aspect ratio of the GNPs, as well as dispersion and morphology, has a strong influence on resulting properties of nanocomposites. For that reason, it is necessary to analyze the morphology of the composites in order to determine the dispersion degree, which is the relative position of the GNPs through the epoxy matrix, in addition to the spatial distribution, referring to the possible exfoliation of graphite nanosheets or the appearance of wrinkling or stretching phenomena [15]. In a previous study carried out by X-Ray diffraction [13], it was confirmed that the exfoliation phenomenon does not occur with the applied dispersion technique.

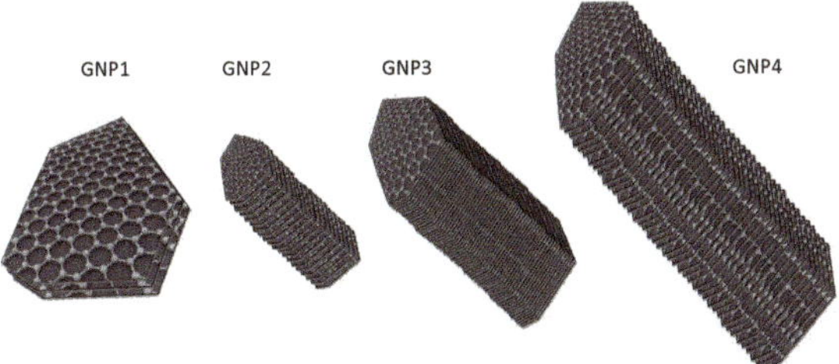

Figure 2. Scheme of graphene nanoplatelets: GNP1, GNP2, GNP3 and GNP4. These schemes are only comparatives. The thickness of nanoplatelets (~nm) is much lower than the lateral dimensions (~µm).

Figure 3 shows micrographs of the studied composites captured by FEG-SEM. These micrographs confirm a suitable dispersion degree for all the studied samples (Figure 3a,b) but also that high shear stirring could induce a wrinkling effect of the nanosheets [13]. This phenomenon seems to be more noticeable for composites reinforced with lower thickness GNPs (Figure 3c,d), while GNPs with higher thickness remain mostly stretched (Figure 3e,f). It is worthy to note that the percentage of GNPs added was the same one for all the samples. However, making a comparison between the micrographs (Figure 3a,b), it is possible to observe that the concentration of nanoplatelets, determined as the number of GNPs per observed area, is highest in the epoxy resin reinforced with GNP1. This concentration seems to decrease as the size of the nanoplatelets does. This is due to the different aspect ratios of the GNPs. GNP1 GNPs have the highest specific surface area and lowest thickness whereas GNP4 GNPs have specific surface area several orders of magnitude lower than GNP1 ones and higher thickness.

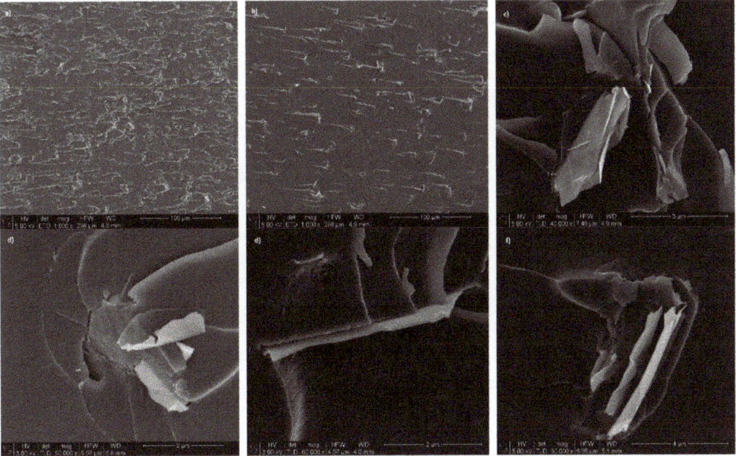

Figure 3. FEG-SEM micrographs of epoxy composites reinforced with GNP1 (**a**,**c**); GNP 2 (**d**), GNP3 (**e**) and GNP4 (**b**,**f**).

3.2. Moisture Absorption

Figure 4 shows the moisture absorption curves obtained for neat epoxy resin and composites reinforced with the different types of graphene nanoparticles. Firstly, it is worth pointing out that the pristine epoxy resin studied in this work is aircraft quality and, therefore, its hydrothermal resistance is initially very high, absorbing only 1.75% w/w of water. For this reason, the enhancement reached by graphene addition is less pronounced than other published works with epoxy resins of different nature, with a curing process at room temperature [5,6]. The crosslinking degree of these resins is lower and thus, their water uptake was higher, close to 6% w/w.

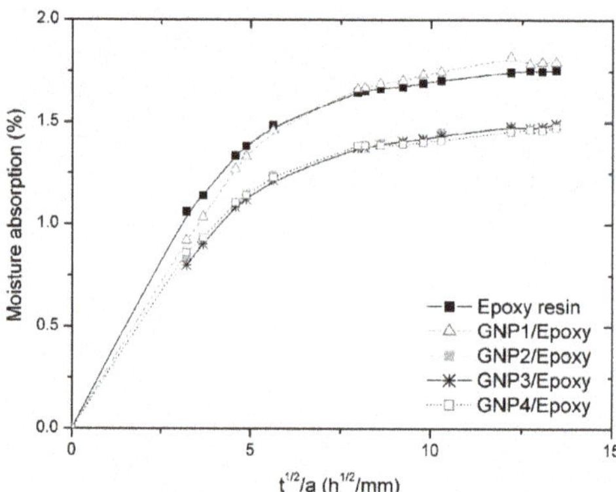

Figure 4. Moisture absorption of neat epoxy resin and epoxy composites reinforced with different types of graphene, commercially named GNP1, GNP2, GNP3 and GNP4, were t is time, and a is area.

At the first stage, the water uptake curves present a tendency that can be approximated to linear growth. After this first region, the tendency changes and the curves asymptotically approach the equilibrium. From these results, the diffusion coefficient can be calculated by knowing the water absortion rate, according to the Fick´s model from the initial slope of the absorption curves [16]. The other characteristic parameter is the maximum absorbed water content when the saturation is reached. These parameters are summarized in Table 2 for all the studied samples. It is worthy to note that the incorporation of graphene nanoplatelets into the epoxy matrix causes different effects on the water uptake of the resin as a function of the GNPs geometry.

Table 2. Water diffusion coefficients (D) calculated by Fick model and weight fraction of water at the saturation equilibrium (W∞).

Sample	$D \times 10^9$ (cm$^2 \cdot$s^{-1})	W∞ (%)
Epoxy resin	10.8	1.75
GNP1/epoxy	7.99	1.79
GNP2/epoxy	7.97	1.78
GNP3/epoxy	6.56	1.46
GNP4/epoxy	7.53	1.45

In fact, it is striking that the addition of GNPs induces a reduction of the maximum absorbed moisture content in most cases except for GNP1/epoxy composites. However, this is the nanofiller with the highest specific area and aspect ratio, therefore, it would be expected that this composite presents the highest barrier properties. These controversial results can be explained by the morphology of the composites. As it is shown in Figure 3d, GNP1 suffers an important phenomenon of weaving. This implies two different effects: A decrease of its effective area and a weak interphase with the matrix, that results in areas with low wettability. Despite GNPs difficult water absorption, the final amount of absorbed water is higher due to this weaker interface, which allows water to go in between. In contrast, the addition of only a 0.5 wt % of GNP2, GNP3 and GNP4 causes an important reduction of water absorption. It confirms that, when the graphite nanosheets are stretched, they act as effective barriers. This affirmation is corroborated by the diffusion coefficient. The water absorption rate decreases by the GNPs addition. This parameter is lower for epoxy/graphene composites with regards to neat epoxy resin, but this remains practically constant for all the studied composites. This decrease of diffusion coefficient by graphitic nanofiller addition has been already observed by other authors.

As it is indicated by O. Starkova et al. [5,6], water absorption is a complex phenomenon, influenced by many factors, including free volume, crosslinking degree of the matrix, morphology, hydrophobicity, etc. [17]. They explain the increase of water resistance in epoxy composites, reinforced with nanofillers, by two effects. In the first place, nanofillers act as efficient barriers against water absorption due to the increased tortuosity for water molecules diffusing through the epoxy matrix. The second reason is that nano-sized particles restrict intermolecular movements of the surrounding epoxy thus, retarding the relaxation of polymer chain segments. This would explain the results in the present work. Graphene addition induces a decrease of the absorbed water content and diffusion coefficient. A higher specific surface area enhances the barrier properties when graphene nanoplatelets are well dispersed and stretched. Nevertheless, the thinnest nanosheets (GNP1) can suffer wrinkling during the dispersion stage, thus the molecular mobility restriction is less effective, increasing the maximum absorbed water content.

3.3. Thermal Effect

The effect of the water absorption in the epoxy resin and composites was studied by DMTA. DMTA specimens were introduced in the climatic chamber and removed at selected periodic times. In all cases, the curves of loss of tangent (tan δ) presented only one peak. A shoulder at low temperature was not observed in any case, this behavior has also been observed by other authors [5,6]. The reason is the low moisture content absorbed by the studied system. Differences in T_g due to duplicate samples did not exceed 1–2 °C, while the standard deviation obtained in the storage modulus in the glassy state was higher, close to 10%. For this reason, the effect of moisture uptake on the mechanical properties was studied by flexural test. In spite of the good reproducibility of T_g measurements, the T_g decreases due to the fact that the plasticization effect was small, in account of the low water absorption content. For this reason, the T_g was measured at different aging times in order to obtain the tendency, which is shown in Figure 5.

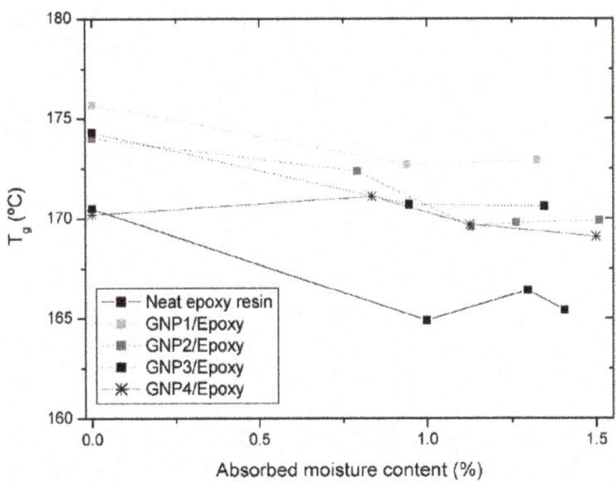

Figure 5. Decrease of glass transition temperature due to the water absorption.

As produced, the glass transition temperature increases by the GNPs addition in a range of 4 or 5 °C, except for the composites reinforced with GNP4, whose T_g remains constant. Again, it is worthy to note that the initial T_g value of pristine resin is very high due to its high crosslinking degree, associated to its high curing temperature (140 °C). This could be associated to the rheological percolation [5,17]. Nanofiller loads close to percolation hinder the mobility of polymer chain segments. The percolation threshold decreases with the increase of the specific surface area of nanofillers. GNP4 has lower specific surface area and higher thickness (Table 1), which implies that individual nanoparticles, due to their higher volume, are not as close as they are when using GNP1, GNP2 or GNP3 (see Figure 3c), limiting its influence on the polymer chain mobility and making glass transition temperature remain constant.

The moisture absorption induces a T_g decrease. This is the best-known consequence of the water plasticization. As was expected, the T_g decrease is more noticeable at the beginning of the aging treatment and, after that, T_g remains nearly constant. This could be related to a catalytic effect. Absorbed water molecules can form hydrogen bonds (named Type II or bonded water) with some remaining oxirane groups, enhancing a postcuring reaction. In fact, this phenomenon could explain the dispersion of glassy storage modulus values. On the other hand, the plasticization effect of water should decrease the rigidity of the matrix, enhancing the chain mobility, but the resultant postcuring reaction would increase the crosslinking degree. Each of them provokes contrary effects on the storage modulus.

The T_g of neat epoxy resin decreases 3.6 °C per each 1% in weight of moisture absorbed while this effect decreases in composites up to 2.5 °C. This means that the addition of graphene nanoplatelets shows numerous advantages related to the hydrothermal resistance: Decreasing the absorbed water content and the consequences of its plasticization.

3.4. Mechanical Effect

Finally, as mentioned above, the modification of mechanical properties due to hydrothermal aging was also studied by DMTA. Table 3 collects the results obtained for epoxy resin and composites reinforced with different types of graphene at different aging states. The specimens were aged during 1 and 17 days, respectively.

Table 3. Mechanical properties of neat epoxy resin and composites at different aging states.

	Moisture Absorption (%)	E (GPa)	σ (MPa)	ε (%)
Epoxy resin	0	2.24 ± 0.24	146 ± 20	5.7 ± 1.3
	0.86	2.32 ± 0.02	137 ± 7	4.5 ± 0.2
	1.55	3.61 ± 0.44	166 ± 29	3.5 ± 0.3
GNP1/epoxy	0	4.20 ± 0.37	140 ± 10	1.9 ± 0.5
	0.92	3.94 ± 0.02	120 ± 2	2.1 ± 0.1
	1.79	3.87 ± 0.23	95 ± 3	1.9 ± 0.1
GNP2/epoxy	0	4.30 ± 0.80	111 ± 4	2.6 ± 0.4
	0.84	3.41 ± 0.05	133 ± 12	3.2 ± 0.4
	1.49	3.39 ± 0.09	127 ± 20	2.9 ± 0.3
GNP3/epoxy	0	3.52 ± 0.53	97 ± 17	2.0 ± 0.1
	0.80	3.48 ± 0.03	113 ± 2	2.4 ± 0.1
	1.49	3.42 ± 0.51	101 ± 1	2.3 ± 0.4
GNP4/epoxy	0	3.33 ± 0.0	136 ± 5	2.7 ± 0.1
	0.85	3.09 ± 0.05	93 ± 4	2.9 ± 0.2
	1.47	3.18 ± 0.16	137 ± 18	3.3 ± 0.3

The water uptake on pristine resin induces an important increase of stiffness, a slight increase of mechanical strength and a decrease of elongation. These phenomena confirm that the postcuring reaction is catalyzed by the absorbed water, increasing the crosslinking degree. However, this effect is scarcely observed in composites. The stiffness and mechanical properties remain constant or suffer slight decreases by the moisture absorption. The decrease of mechanical properties could be associated to two different causes. Firstly, the water induces plasticization and also promotes the weak GNP-matrix interphase. Secondly, the slight increase of elongation on graphene/epoxy composite is clearly explained by the plasticization effect of the matrix.

Initially, the mechanical behavior of non-aged samples is analyzed. The graphene addition induces an important increase of stiffness in spite of the high elastic modulus of the pristine resin. The addition of 0.5% GNPs induces an increase of the elastic modulus from 2.24 GPa to 3.94 GPa (76%) for GNP1/epoxy composites. The modulus increase of the composites seems to be proportional to the specific surface area of the graphene nanoplatelets, being the highest for GNP1 (76%) and the lowest for GNP3 and GNP4 (41%). In contrast, the mechanical strength and elongation at break markedly decrease with the graphene addition. This is mainly associated with the weak interphase, which makes necessary the incorporation of functional groups in the nanofiller in order to enhance a chemical interphase with the epoxy matrix [17,18].

In contrast to the thermomechanical properties, the influence of moisture in mechanical properties of the neat epoxy resin and nanocomposites differ. In order to analyze it in depth, Figure 6 shows the variation of each flexural property as a function of the amount of absorbed moisture for neat epoxy resin and graphene/epoxy composites. In general, at the same moisture content, the mechanical properties of epoxy resin are significantly influenced by the presence of water in the network while the ones of composites are slightly influenced. Similar behavior was observed in the water uptake effect on the glass transition temperature. As mentioned above, the addition of graphene nanofillers induces a molecular mobility restriction, reducing the water uptake. This hindering of intermolecular movements could also cause the water effect to be lower in the composites. One issue to take into account is that, due to the presence of microcavities between nanoplatelets and the epoxy matrix, water molecules can diffuse and locate at the interphase of the nanoplatelets. This preferential location induces a lower effectiveness in the load transfer [19], which is the reason for the differences observed.

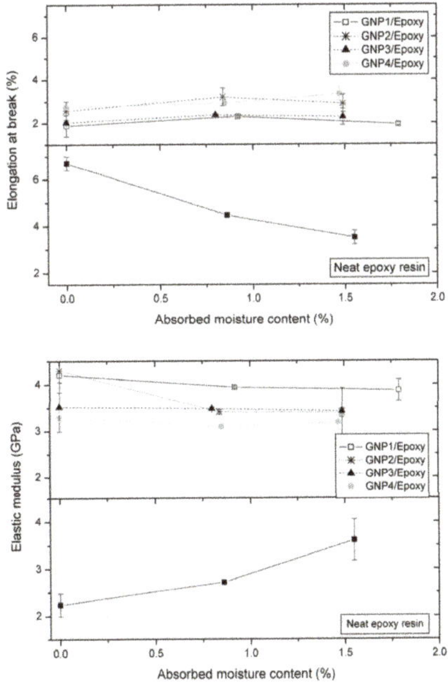

Figure 6. *Cont.*

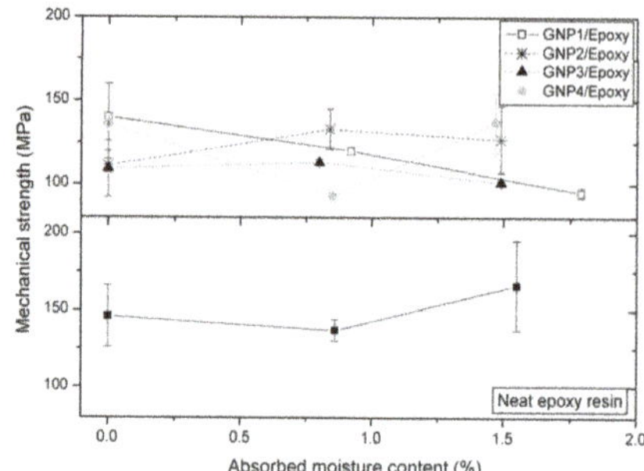

Figure 6. Variation of mechanical properties due to the water absorption.

4. Conclusions

The barrier properties of graphene nanoplatelets, used as nanoreinforcement into epoxy composites, strongly depend on their geometry: Their thickness and lateral dimensions. In general, an increase of the specific surface area induces an increase of hydrothermal resistance of the composite when the nanofiller dispersion is suitable. Thinner nanoplatelets usually are not totally stretched and thus, remaining partially wrinkled, decreasing their efficiency as water barrier elements. The addition of GNPs into epoxy resins has several advantages against moisture absorption: A decrease of the maximum water content, the diffusion coefficient and even the mentioned catalytic effect of absorbed water in which hydrogen bonds are formed. In fact, mechanical and thermal properties suffer variation of less significance by the water entrance in the epoxy network than those of pristine resin.

The addition of GNPs to the epoxy resin reduces the common phenomenon observed in pristine epoxy resin, mainly the plasticization and catalyst effect of postcuring reaction due to the formation of hydrogen bonds between water molecules and residual oxirane rings. This effect has been attributed to mobility restriction of molecular chain segments due to the presence of the nanofillers.

Author Contributions: S.G.P. contributed to article writing, funding support, design of experiments and analysis of results. A.J.-S. contributed to article reviewing, design of experiments, test performing and analysis of results. R.M. contributed to article reviewing, design of experiments, test performing and analysis of results. A.U. contributed to funding support for this research and research supervision.

Funding: This research was funded by Ministerio de Economia y Competitividad of Spain Government (Project MAT2016-78825-C2-1-R) and Comunidad de Madrid (Project P2013/MIT-2862).

Acknowledgments: The authors wish to thank the Ministerio de Economia y Competitividad of Spain Government (Project MAT2016-78825-C2-1-R) and Comunidad de Madrid (Project P2013/MIT-2862).

Conflicts of Interest: The authors declare no conflict of interest.

References

1. Saravanan, N.; Rajasekar, R.; Mahalakshmi, S.; Sathishkumar, T.P.; Sasikumar, K.S.K.; Sahoo, S. Graphene and modified graphene-based polymer nanocomposites—A review. *J. Reinf. Plast. Compos.* **2014**, *33*, 1158–1170. [CrossRef]
2. Mohd Zulfli, N.H.; Abu Bakar, A.; Chow, W.S. Mechanical and water absorption behaviors of carbon nanotube reinforced epoxy/glass fiber laminates. *J. Reinf. Plast. Compos.* **2013**, *32*, 1715–1721. [CrossRef]
3. Kingston, C.; Zepp, R.; Andrady, R.; Boverhof, D.; Fehir, R.; Hawkins, D.; Roberts, J.; Sayre, P.; Shelton, B.; Sultan, Y.; et al. Release characteristics of selected carbon nanotube polymer composites. *Carbon* **2014**, *68*, 33–57. [CrossRef]
4. Guadagno, L.; De Vivo, B.; Di Bartolomeo, A.; Lamberti, P.; Sorrentino, A.; Tucci, V.; Vertuccio, L.; Vittoria, V. Effect of functionalization on the thermo-mechanical and electrical behavior of multi-wall carbon nanotube/epoxy composites. *Carbon* **2011**, *49*, 1919–1930. [CrossRef]

5. Satarkova, O.; Chandrasekaran, S.; Prado, L.A.S.A.; Tölle, F.; Mülhaupt, R.; Schulte, K. Hydrothermally resistant thermally reduced graphene oxide and multi-wall carbon nanotube based epoxy nanocomposites. *Polym. Degrad. Stab.* **2013**, *98*, 519–526. [CrossRef]

6. Satarkova, O.; Buschhorn, S.T.; Mannov, E.; Schulte, K.; Aniskevich, A. Water transport in epoxy/MWCNT composites. *Eur. Polym. J.* **2013**, *49*, 2138–2148. [CrossRef]

7. Liu, W.; Hoa, S.V.; Pugh, M. Water uptake of epoxy–clay nanocomposites: Model development. *Compos. Sci. Technol.* **2008**, *68*, 156–163. [CrossRef]

8. Guadagno, L.; Vertuccio, L.; Sorretino, A.; Raimondo, M.; Naddeo, C.; Vitoria, V. Mechanical and barrier properties of epoxy resin filled with multiwalled carbon nanotubes. *Carbon* **2009**, *47*, 2419–2430. [CrossRef]

9. Prolongo, S.G.; Gude, M.R.; Ureña, A. Water uptake of epoxy composites reinforced with carbon nanofillers. *Compos. Part A* **2012**, *43*, 2169–2175. [CrossRef]

10. Alamri, H.; Low, I.M. Effect of water absorption on the mechanical properties of nano-filler reinforced epoxy nanocomposites. *Mater. Des.* **2012**, *42*, 214–222. [CrossRef]

11. Perrin, F.X.; Nguyen, M.H.; Vernet, J.L. Water transport in epoxy–aliphatic amine networks – Influence of curing cycles. *Eur. Polym. J.* **2009**, *45*, 1524–1534. [CrossRef]

12. DeNève, B.; Shanahan, M.E.R. Water absorption by an epoxy resin and its effect on the mechanical properties and infra-red spectra. *Polymer* **1993**, *34*, 5099–5105. [CrossRef]

13. Moriche, R.; Prolongo, S.G.; Sánchez, M.; Jiménez-Suárez, A.; Sayagués, M.J.; Ureña, A. Morphological changes on graphene nanoplatelets induced during dispersion into an epoxy resin by different methods. *Compos. Part B* **2015**, *72*, 199–205. [CrossRef]

14. Prolongo, S.G.; Jiménez-Suárez, A.; Moriche, R.; Ureña, A. Graphene nanoplatelets thickness and lateral size influence on the morphology and behavior of epoxy composites. *Eur. Polym. J.* **2014**, *53*, 292–301. [CrossRef]

15. Liu, S.; Yan, H.; Fang, Z.; Wang, H. Effect of graphene nanosheets on morphology, thermal stability and flame retardancy of epoxy resin. *Compos. Sci. Technol.* **2014**, *90*, 40–47. [CrossRef]

16. Crank, J. *The Mathematic of Diffusion*; Clarendon Press: Oxford, UK, 1956.

17. Choudalakis, G.; Gotsis, A.D. Free volume and mass transport in polymer nanocomposites. *Curr. Opin. Colloid Interface Sci.* **2012**, *17*, 132–140. [CrossRef]

18. Ahmadi-Moghadam, B.; Sharafimasooleh, M.; Shadlou, S.; Taheri, F. Effect of functionalization of graphene nanoplatelets on the mechanical response of graphene/epoxy composites. *Mater. Des.* **2015**, *66*, 142–149. [CrossRef]

19. Tam, L.; Wu, C. Molecular Mechanics of the Moisture Effect on Epoxy/Carbon Nanotube Nanocomposites. *Nanomaterials (Basel)* **2017**, *7*, 324. [CrossRef] [PubMed]

Article

Thermal Properties of PEG/Graphene Nanoplatelets (GNPs) Composite Phase Change Materials with Enhanced Thermal Conductivity and Photo-Thermal Performance

Lihong He [1,2,*], Hao Wang [1,2,*], Hongzhou Zhu [2], Yu Gu [3], Xiaoyan Li [1] and Xinbo Mao [1]

[1] School of Materials Science and Engineering, Chongqing Jiaotong University, Chongqing 400074, China; 990201400027@cqjtu.edu.cn (X.L.); maoxinbo_19930303@163.com (X.M.)

[2] The National Joint Engineering Laboratories of Traffic Civil Materials, Chongqing Jiaotong University, Chongqing 400074, China; zhuhongzhouchina@126.com

[3] The Key Laboratory of Road and Traffic Engineering, Ministry of Education, Tongji University, No. 4800 Cao'an Road, Shanghai 201804, China; guyu8241183@163.com

* Correspondence: sunnyhlh@126.com (L.H.); SodaWangHao@163.com (H.W.); Tel.: +86-138-9602-3469 (L.H.); +86-132-0616-7705 (H.W.)

Received: 3 November 2018; Accepted: 9 December 2018; Published: 13 December 2018

Abstract: This paper mainly concentrates on the thermal conductivity and photo-thermal conversion performance of polyethylene glycol (PEG)/graphene nanoplatelets (GNPs) composite phase change materials (PCMs). The temperature-assisted solution blending method is used to prepare PCM with different mass fraction of GNPs. According to the scanning electron microscope (SEM), GNPs are evenly distributed in the PEG matrix, forming a thermal conduction pathway. The Fourier transform infrared spectra (FT-IR) and X-ray diffraction (XRD) results show that the composites can still inherit the crystallization structure of PEG, moreover, there are only physical reactions between PEG and GNPs rather than chemical reactions. Differential scanning calorimeter (DSC) and thermal conductivity analysis results indicate that it may be beneficial to add a low loading ration of GNPs to obtain the suitable latent heat as well as enhance the thermal conductivity of composites. To investigate the change in the rheological behavior due to the effect of GNPs, the viscosity of the composites was measured as well. The photo-thermal energy conversion experiment indicates that the PEG/GNPs composites show better performance in photothermal energy conversion, moreover, the Ultraviolet-visible-Near Infrared spectroscopy is applied to illustrate the reasons for the higher absorption efficiency of PEG/GNPs for solar irradiation.

Keywords: polyethylene glycol; phase change materials; graphene nanoplates; thermal conductivity; photo-thermal conversion performance

1. Introduction

The rapid consumption of fossil fuels and the increasing contradiction between energy supply and demand are compelling researchers to utilize energy more effectively and develop renewable energy [1]. A solar energy source, which is inexhaustible, economical and environmentally friendly, has been widely recognized as an ideal form of renewable energy [2]. In recent years, the conversion of photo-thermal technology is the most successful and popular technique among the different forms of solar energy utilization, such as light-thermal conversion, light-electricity conversion, light-biology conversion [3]. Photo-thermal applications not only possess the merits of relatively high photo-thermal energy conversion efficiency but also can be achieved without complicated and expensive instruments [4,5]. However, there are still some nonnegligible drawbacks in the use of solar energy, such as low conversion efficiency, diurnal fluctuation of optical radiation, which limit the efficient utilization of solar thermal energy storage [4]. Latent heat thermal energy storage (LHTES) system has proved to be a promising technique for overcoming the drawbacks of solar energy conservation because of its outstanding advantages, such as high heat storage density, constant operating temperature, isothermal characteristics. In recent years, LHTES has been widely applied to various thermal storage management applications [6–9]. Phase change materials (PCMs), as representative of advanced LHTES materials, are the most prevalent and effective technique for thermal energy storage because of their high enthalpy change, non-toxic, and reusability characteristic, which can be utilized in a range of applications, such as thermal management of solar energy, insulation clothing, thermal insulation buildings, spacecraft thermal control and so on [10–14].

Among various organic PCMs (OPCMs), polyethylene glycol (PEG) is universally recognized as a kind of outstanding solid-liquid transformation PCM because of its excellent properties, including economically available, high phase change enthalpy, biodegradation, little subcooling, low vapor pressure, non-toxicity [14–16]. In addition, it is convenient to obtain the suitable phase change temperature and enthalpy via simply changing molecular weight of PEG [17]. However, the pristine PEG as OPCMs in the utilization of solar energy has intrinsic shortcomings, including low thermal conductivity and poor absorptive performance in the optical light, which accounts for 50% of the solar radiation energy. The former refers to the speed of absorbing and releasing thermal energy, which can reduce the energy efficiency [18]. The latter leads to low solar energy efficiency, which limits its applications of solar thermal energy to a great extent. Consequently, the thermal conductivity and photothermal performance of OPCM are imperative to be enhanced.

Hence, many prominent works have been done to obtain the PCMs with much higher thermal conductivity. A prevalent approach against this shortcoming is to incorporate the highly thermal conductive fillers into the PCMs, such as silver nanowire [19], Cu [20], TiO_2 nanoparticles [21]. Deng et al. [19] achieved PEG-Ag/expanded vermiculite PCMs by the physical blending and impregnation method. The thermal conductivity increased to 0.68 W/(mK) for 19.3 wt % silver nanowire in the composite PCMs, which was 11.3 times higher than that of pristine PEG. Zhang et al. [20] prepared composite PCMs by adding Cu powder to PEG/SiO_2 through the sol-gel method, and discovered that the thermal conductivity of the composite PCMs reached up to 0.431 W/(mK) by an addition of 3.45 wt % Cu powder, which was enhanced by 49.13% in comparison with pure PEG. Harikrishnan et al. [21] used a two-step method to prepare stearic acid-TiO_2 nanofluids PCMs. For composite with only 0.3 wt % TiO_2, the enhancement of thermal conductivity was measured to be 70.52%. However, the aforementioned composite PCMs are white powders, which present poor absorptive performance in the optical light, resulting in a weak photothermal conversion property and a low solar energy utilization ratio. Therefore, the carbon materials are good additives for improving the optical absorption performance of PCMs because of their high visible light absorptivity. Graphene nanoplates (GNPs) are thin flat particles that consist of single and few layer graphene mixed with thicker graphite, thus, structurally they are in between graphene and graphite [22]. Compared to other carbon-based materials, GNPs possess many outstanding functional performance, such as unique photonic/optical transportation, light weight, high thermal/electrical conductivity, excellent mechanical property [23,24]. Furthermore, in the optical regime, the GNPs have only very limited loss at Dirac point because of their outstanding optical characteristics, including ballistic transport and saturable absorption [25–28]. Therefore, GNPs can be easily and successfully incorporated with polymeric matrices to enhance the comprehensive performance of polymers. As for PCMs, it has already been demonstrated that the addition of even a small mass fraction of GNPs can significantly improve the thermal conductivity of different PCMs such as palmitic acid [23,29], beeswax [12], eicosane [30], 1-octadecanol [31], decosane [32]. In addition, the rheological behavior of PCMs can be significantly influenced by the addition of nanoplates, which determines the transportablilty in pump systems for PCMs nanofliuds applications. However, in previous literatures, the comprehensive effects of GNPs on the thermal conductivity, rheological behavior, and optical absorption performance of PCMs are hardly explored together, hence, this work will focus on analyzing the three effects.

In this paper, a series of PEG/GNPs composites were prepared through a temperature-assisted solution blending method in order to develop new PCMs for utilizing solar energy. The good compatibility between PEG and GNPs conduce to excellent thermal conductivity and high latent heat at low filler mass fraction. Furthermore, the composites exhibit a better performance in absorbing and conversing solar energy in comparison with the traditional organic PCMs. Therefore, the outstanding thermal conductivity and photothermal performance make the PEG/GNPs composites a promising candidate for solar thermal energy storage application.

2. Experimental

2.1. Materials

PEG (Mn = 4000), as the latent heat storage material, was purchased from Chengdu Kelong Chemical Reagent Factory (Chengdu, China). GNP nanoplates (Grade C750) (Thickness: <2 nm, Width: <2 µm, Specific surface area: 750 m^2/g, purity: >99%) were purchased from XG Science (Lansing, MI, USA). Ethanol was obtained from Chongqing Chuandong Chemical Group (Chongqing, China). All the materials were of analytical grade.

2.2. Preparation of the PEG/GNPs Composites

PEG/GNPs composites were prepared by the temperature-assisted solution blending method. Firstly, GNPs were dispersed in ethanol in aid of ultrasonication for 1 h to form a homogeneous suspension, meanwhile, the PEG was heated by water bath at 80 °C in a beaker. Then the GNPs suspension was slowly dropped into melted liquid PEG and the hybrid solution was stirred vigorously for 4 h under the same conditions to evaporate the remaining ethanol. Finally, the products were dried in a vacuum oven to a constant weight at 60 °C. For comparison purposes, pristine PEG was also prepared with the same procedure. Here, the mass fraction of GNPs in PEG/GNPs composites varied from 0.5 to 2% (0.5, 1, 1.5 and 2 wt %), for convenience, the obtained samples were labeled as PEG/GNPs-X%, where X was characteristic of the GNPs mass fraction in the composites.

2.3. Characterization

A EVO18 scanning electron microscope (SEM) instrument (CARL ZEISS, Oberkochen, Germany) was used to visually characterize the morphology of the composites with an accelerating voltage of 20 kV. All samples were sputtered with gold prior to test.

Fourier transform infrared (FT-IR) spectra of PEG, GNPs, and PEG/GNPs composites were performed on a Nicolet 6700 (Nicolet Instrument Company, Waltham, MA, USA) at the wavenumber range of 4000–500 cm^{-1}.

The crystal structures and crystallization characteristic of PEG in composites was explored by X'Pert Powder model of X-ray diffractometer (PANalytical B.V., Almelo, the Netherlands) with Cu Kα radiation (λ = 0.154 nm) under a voltage of 40 kV and a current of 40 mA. The scanning angle 2θ, from 10° to 40° at a scanning speed of 3°/min.

The thermal energy storage properties of the composites were investigated using a differential scanning calorimeter (DSC) (DSC6000, Perkin-Elmer Inc., Waltham, Mass, USA) at a heating and cooling rate of 10°/min under a constant stream of nitrogen. This process was repeated for the three heating-cooling cycles and the third one was used to analyze the result.

The UV-VIS-NIR spectroscopy was performed on a UV-3600 spectrophotometer (SHIMADZU, Tokyo, Japan).

The thermal conductivity of the samples was measured using a LFA 447 MicroFlash Apparatus (Netzsch, SELB, Germany) by the Laser Flash method. To ensure the accuracy of the measurement, the thermal conductivity of each sample was measured five times.

The viscosity of the composites was tested using a viscometer (DVDV-I, Brookfield, Middleboro, MA, USA) with accuracy within ±1%.

The light-to-thermal energy conversion experiment was conducted using a 300 W solar simulator (ULTRA-VITALUX, OSRAM, Munich, Germany) as the light source. During the test, the temperatures of the samples were recorded using a L93-6 temperature logger with thermocouples (Hangzhou Loggertech Co., Ltd., Zhejiang, China).

3. Result and Discussion

3.1. Microstructure Analysis

The microstructures of the GNPs, pristine PEG and PEG/GNPs-2% composite are shown in Figure 1. The GNPs in Figure 1a exhibit wrinkled surface textures with curling edges, and they are especially prone to agglomerate because of their high specific surface areas and strong π-π interaction. Actually, these wrinkled surface with a lot of creases can play a positive role in enhancing the strong interaction between GNPs with PEG. For pristine PEG, a relatively smooth surface and compact structure appears (Figure 1b), completely different from that of GNPs. The morphology of the PEG/GNPs composite is shown in Figure 1c, as can be seen, the surface of PEG displays a conspicuous change that appears to be rougher than that of pure PEG, which may be attributed to the uniform dispersion of GNPs in the PEG matrix. The composite presents a coarse surface with plenty of creases stacking layer by layer, which seems to provide a network-like structure that the phonons can efficiently travel along and accelerate heat transfer, thus, there will be a significant enhancement in the thermal conductivity of composites in comparison with that of pure PEG [31]. Moreover, Figure 1b,c confirm the suitable interactions and desirable compatibility between GNPs and PEG matrix because of the van der Waals force, hydrophobic-hydrophobic interaction and π-π interaction [24,33]. Therefore, it can be demonstrated that the GNPs can be used as a kind of filler material for the fabrication of PEG with remarkable thermal conductivity at a low loading ration.

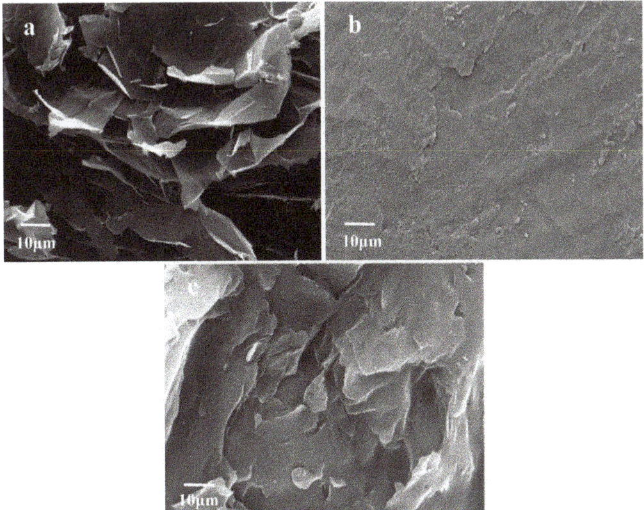

Figure 1. Scanning electron microscope (SEM) images of the graphene nanoplatelets (GNPs) (**a**), pristine polyethylene glycol (PEG) (**b**) and PEG/GNPs-2% composite (**c**).

3.2. FT-IR Analysis

The chemical compatibility of the PEG/GNPs composites was characterized by FT-IR spectroscopy (Figure 2). As shown in Figure 2, in the spectrum of pure PEG, the absorption peak at 3435 cm^{-1} belongs to the stretching vibration of O–H groups while the sharp peak at 2881 cm^{-1} is attributed to the C–H stretching vibration. Moreover, the obtained bands at 1642 cm^{-1} and 1110 cm^{-1} represent the C=O stretching vibration and C–O asymmetric stretching vibration, respectively. In addition, the C–H bending vibration is shown at 841 cm^{-1}. Similar observations were reported in a previous study [34]. In terms of the spectrum of GNPs, it appears nearly a flat line because it is universally acknowledged that there are few functional groups on GNPs [18]. It can be clearly seen that the absorption peaks in the spectrum of PEG/GNPs composites are in accordance with the spectrum of PEG, which reveals that the previously mentioned peaks of PEG remain constant in the composites. Furthermore, no distinct new peaks are observed in the spectra of the mixtures, suggesting that there are only physical reactions between PEG and GNPs rather than chemical reactions.

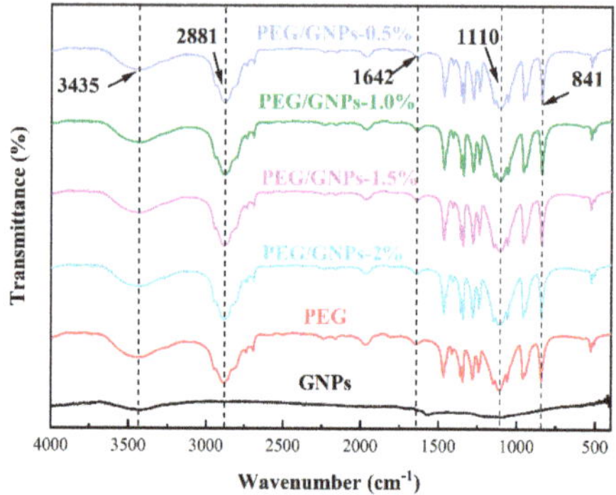

Figure 2. Fourier transform infrared (FT-IR) spectra of GNPs, pure PEG and PEG/GNPs composites with different contents of GNPs.

3.3. X-ray Diffraction (XRD) Analysis

Figure 3 presents the XRD patterns of the pure PEG, GNPs and PEG/GNPs composites. The low intensity diffraction peak at 26.23° is the main peak of GNPs, which reveals the random stacking of a few of graphene sheets and represents the crystallization of GNPs [23]. The two sharp diffraction peaks of the PEG appeared at 19.06° and 23.17°, which indicates a polymer with high crystallinity. After GNPs are uniformly dispersed into the PEG matrix, the typical diffraction peaks of PEG can still be observed and the peak positions do not change. However, the characteristic peak of the GNPs is not shown in the PEG/GNPs composites, which can be ascribed to the fact that the GNPs loading is extremely low compared to that of PEG. One the other hand, the peak intensities of PEG slightly changed after the incorporation of GNPs, especially for the peak at 23.17°. For the composites, the intensities of peaks at 23.17° are relatively lower than that of pristine PEG when the content of GNPs varies from 0.5% to 1.5%. The possible reason is that the strong interaction between GNPs and PEG can restrict the mobility of PEG molecular chains, resulting in the decrease in crystallization of PEG. Therefore, from the XRD results, it can be confirmed that the PEG/GNPs composites can still possess well-maintained crystallization behaviors, and there is no significant effect of the GNPs on the crystal structure of the pure PEG. Moreover, the XRD results can be seen as an indication that no chemical reaction between GNPs and PEG occur.

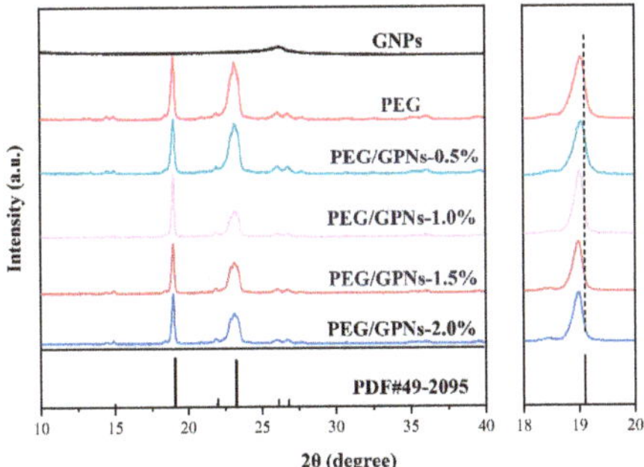

Figure 3. X-ray diffraction (XRD) patterns of GNPs, pure PEG and PEG/GNPs composites with the different contents of GNPs.

3.4. Thermal Storage Performance Analysis

The phase change temperature and thermal energy storage properties of pristine PEG and PEG/GNPs composites were measured by the DSC technique to explore the effect of GNPs on the thermal storage performance of composites. The melting and solidifying DSC curves of PEG and PEG/GNPs composites are shown in Figures 4 and 5, respectively. The detailed calorimetric results of the DSC experiments are tabulated in Table 1, including starting melting/solidifying temperature (T_{ms}/T_{ss}), end melting/solidifying temperature (T_{me}/T_{se}), peak melting/solidifying temperature (T_{mp}/T_{sp}), endothermic/exothermic enthalpy ($\Delta H_m/\Delta H_c$). Obviously, for all samples, the conspicuous endothermic and exothermic peak are both exhibited in the melting and solidifying process, which represent the solid-liquid phase change of the pure PEG. As can be observed from Table 1, compared with pristine PEG, the T_s of the composites is not significantly influenced by the addition of GNPs. Notably, the composites present lower melting temperatures than that of pure PEG, which is possible attributed to the incorporation of GNPs. When the GNPs are uniformly dispersed into the PEG matrix, the intimate interaction between GNPs and PEG, such as surface tension forces, π-π interactions, and capillary forces, will confine the mobility of PEG molecules, resulting in the decline of phase change temperature. Additionally, the thermal conductivity of composites can be enhanced significantly with the addition of GNPs, which leads to a rapid thermal response. Similar results were founded in a previous study [34]. Therefore, it is demonstrated that the decrease in T_m between pure PEG and composites can be ascribed to the comprehensive effects of these factors.

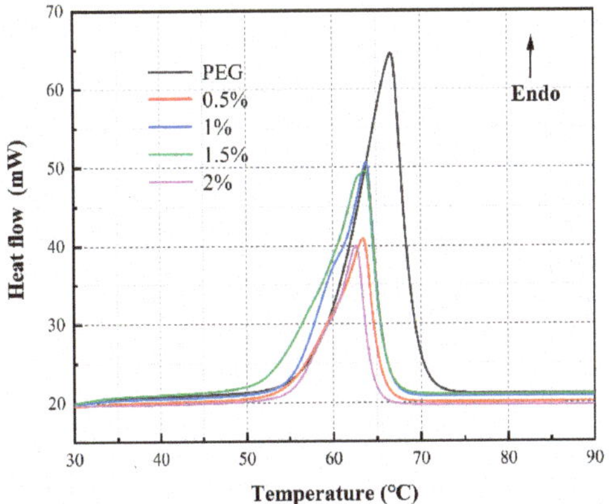

Figure 4. Melting differential scanning calorimeter (DSC) curves of the pure PEG and PEG/GNPs composites with the different contents of GNPs.

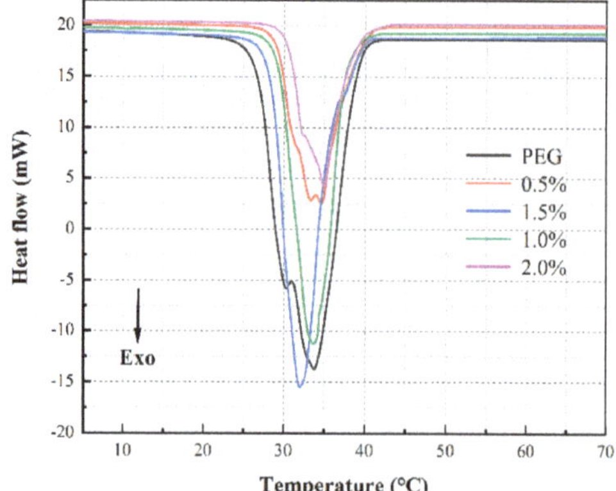

Figure 5. Solidifying DSC curves of the pure PEG and PEG/GNPs composites with the different contents of GNPs.

Table 1. DSC melting and solidifying characteristics of pristine PEG and PEG/GNPs-X% composites.

Sample	T_{ms} (°C)	T_{mp} (°C)	T_{me} (°C)	ΔH_m (J/g)	T_{ss} (°C)	T_{sp} (°C)	T_{se} (°C)	ΔH_s (J/g)
PEG	58.8	66.9	69.1	180.7	38.9	33.8	27.5	161.3
PEG/GNPs-0.5%	57.5	63.8	65.9	178.5	37.6	34.6	29.3	158.4
PEG/GNPs-1%	57.9	63.4	65.7	175.2	36.4	33.7	29.4	155.6
PEG/GNPs-1.5%	56.4	63.2	65.6	172.1	36.5	32.1	28.6	152.8
PEG/GNPs-2%	56.1	62.6	64.8	170.6	37.1	34.2	29.5	151.3

High latent heat is an imperative factor in PCMs, since it is directly related to the capacity of PCMs for energy storage. As can been seen in Table 1, the ΔH_m of the composites decrease with the increasing mass fraction of the GNPs. One reason for this latent heat loss is that some of the PEG volume is replaced by the GNPs, which do not undergo a phase change. Moreover, it is noticeable that the measured latent heat values of the composites in the process of melting are slightly lower than the arithmetically calculated values by the following equation:

$$\Delta H_{\text{nano-composite}} = \mu \times \Delta H_{\text{PEG}} \qquad (1)$$

where $\Delta H_{\text{nano-composites}}$ is the calculated latent heat value of the PEG/GNPs composite, μ and ΔHPEG are the PEG weight percentage of composite and the latent heat of pure PEG, respectively. The measured and calculated latent heat of the composites in the process of melting are compared in Table 2. As can be seen, the relative errors between the measured and calculated values have a tendency to increase with the increasing mass fraction of GNPs. The measured latent heat capacity of the composites doped with 2 wt. % GNPs are 4.4% lower when compared to the calculated values. This possibility is ascribed to the physical interactions between GNPs and PEG, such as van der Waals force and hydrophobic-hydrophobic interaction, which can restrict the mobility of PEG molecular chains during the process of crystallization, as a result, the phase change enthalpy of PEG decreases [34]. We did not increase the GNPs mass fraction beyond 2 wt. % to avoid further decrease in the latent melting heat. The effect of GNPs content to thermal storage performance is shown in Figure 6. As the GNPs content increases, the latent heat gradually decreases while the thermal conductivity increases, implying that the enhancement of thermal conductivity using GNPs will be accompanied by decreased latent heat in the composites. Whereas adding 2 wt. % GNPs results in a 146% increase in thermal conductivity and only a 6.3% reduction in the latent heat of the PEG used in the current work. Therefore, it may be beneficial to add a low loading ration of GNPs to obtain the suitable latent heat as well as enhance the thermal conductivity of composites.

Table 2. Calculated and measured values of the melting latent heat of composites.

Mass Fraction of GNPs (%)	The Calculated Latent Heat (J/g)	The Measured Latent Heat (J/g)	The Relative Errors (%)
0.5	179.8	178.5	0.7
1.0	178.9	175.2	2.1
1.5	178.0	172.1	3.1
2.0	177.1	169.3	4.4

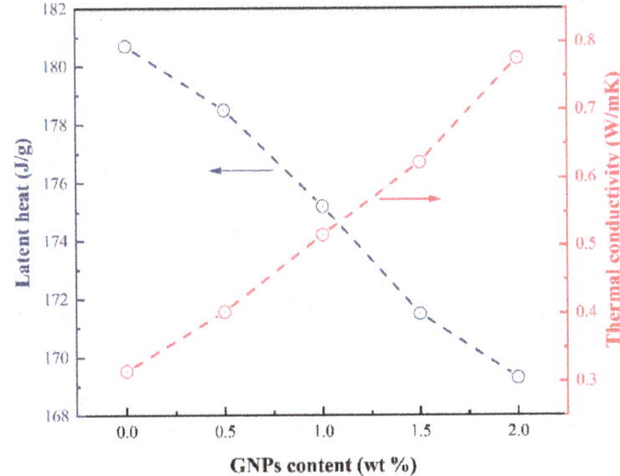

Figure 6. The latent heat and thermal conductivity of the composite as a function of GNPs contents.

3.5. Thermal Conductivity Analysis

The thermal conductivity results of PEG and PEG/GNPs composites are shown in Figure 7. The thermal conductivity of PEG is only 0.316 W/mK, which fails to meet the requirements for efficient thermal storage application. Impressively, it is seen that the thermal conductivities of PEG/GNPs composites increase remarkably as the GNPs weight fraction increases. The thermal conductivity of the composite doped with 2.0 wt % GNPs (PEG/2GNPs) is measured to be as high as 0.776 W/mK, exhibiting a relative increment of above 146%. Moreover, the figure illustrates that the thermal conductivity has a tendency to ascend, implying that GNPs are excellent additives for improving the thermal conductivity of PEG.

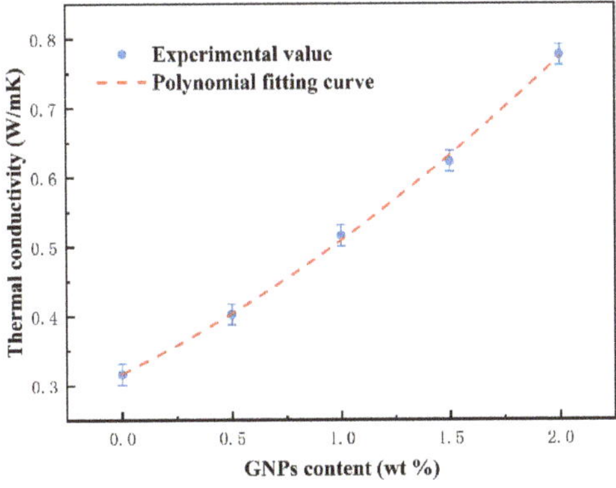

Figure 7. Thermal conductivity of the composites with different contents of GNPs.

GNPs dispersing in PEG matrix can form uniformly aqueous system when the mass fraction of GNPs is relatively low in aid of magnetic stirring and ultrasonic bath. Thus, at a certain filling rate, the different layers of graphene nanoplates interact effectively with each other, forming a high thermal conductivity network. Moreover, the large specific surface area of GNPs greatly reduces the interfacial thermal resistance between the GNPs and PEG matrix materials, as a result, heat will rapidly transfer among GNPs instead of through the PEG matrix. Yu et al. [35] and Fan et al. [36] point out that the low interfacial thermal resistance provided by GNPs is ascribed to the reduced geometric contribution of phonon scattering at the interfaces.

In this work, the polynomial fitting curve is aimed at demonstrating the relationship between thermal conductivity of the PEG/GNPs and the content of GNPs. The dotted red line in Figure 7 represents the polynomial fitting curve, and the relation is expressed as following:

$$\gamma = 0.3168 + 0.156x + 0.036x^2 \tag{2}$$

where γ is thermal conductivity of the composites, x represents the content of GNPs. The coefficient of correlation of the Equation (2) is 0.998, which exhibits a high correlation between the thermal conductivity of composites and the content of GNPs.

High thermal conductivity is a significant element considered for the advanced thermal energy and heat transfer of PCMs. Thus, the enhanced heat transfer rates of PEG/GNPs composites were further investigated. A temperature recorder with thermocouples was used to record the melting and solidifying times for the purpose of evaluating the temperature response behavior of the composites. Before the test, the sample was heated to 90 °C in the 20 mL tube, then the thermocouple was inserted into the middle of the sample, finally, the sample cooled down to room temperature and contacted closely with the thermocouple. The temperature-time profiles of the pristine PEG and composites for both the processes are illustrated in Figure 8. In the melting process, it can be vividly observed that the rising rate of temperature is significantly much faster in PEG/GNPs-2% composite. In addition, the peak temperature of PEG/GNPs-2% reaches 74.9 °C, higher than that of PEG. It is clearly revealed that the GNPs can accelerate the conductivity thermal transfer during the heating process. Then the tubes were immediately put into another thermostatic water baths with the constant temperature of 25 °C. Expectedly, the temperature of pristine PEG continuously decreases at lower rate than that of PEG/GNPs-2% composite due to the different capacity of conductivity thermal transfer. For instance, from the curves depicted in Figure 8, the times taken by pure PEG and PEG/GNPs-2% composite for decreasing to 30 °C from the peak temperatures are determined as about 545 s and 1665 s, respectively, indicating that the composites are more efficient in storing and releasing thermal energy. The faster melting and solidifying rates demonstrate that the thermal conductivity of composites can be significantly promoted by adding the GNPs. From the results mentioned above, it is confirmed that the PEG/GNPs composite has the potential for effective thermal energy management.

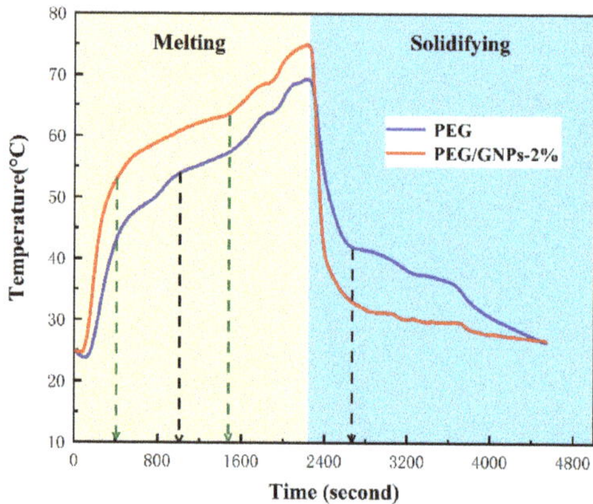

Figure 8. Heating and cooling-rate curves of pure PEG and PEG/GNPs-2% composites.

3.6. Rheological Behavior Analysis

To investigate the rheological behavior of PEG/GNPs composites during the phase change process, a Brookfield DVDV-I viscometer was used to measure the viscosity of the composite fluids at a constant shear stress. The composites were melted before being poured into the sample catcher, and the thermostatic water baths were set to 70 °C to keep the PEG/GNPs composites in the liquid phase. Figure 9 illustrates the viscosity of the composites with different concentrations of GNPs using the ratio method. As can be seen, the viscosity of the composites exhibit a tendency to increase with increasing the mass fraction of GNPs in agreement with other studies [37]. For example, when the loaded GNPs content increases from 0% to 2%, the viscosity of the PEG/GNPs composite fluid increases from 0.11 cP to 0.32 cP at the same temperature, showing a 190.1% enhancement in viscosity compared to the PEG base fluid. It is indicated that the addition of GNPs can increase the viscosity of PEG fluid, which can be attributed to the larger interfacial area of the GNPs with a greater mass fraction in the fluid, and the increase in frictional forces among the GNPs of a larger number [38]. Amin et al. [12] obtained the similar results, showing that the viscosity of PCMs nanofliuids can be enhanced by the addition of nanoplates.

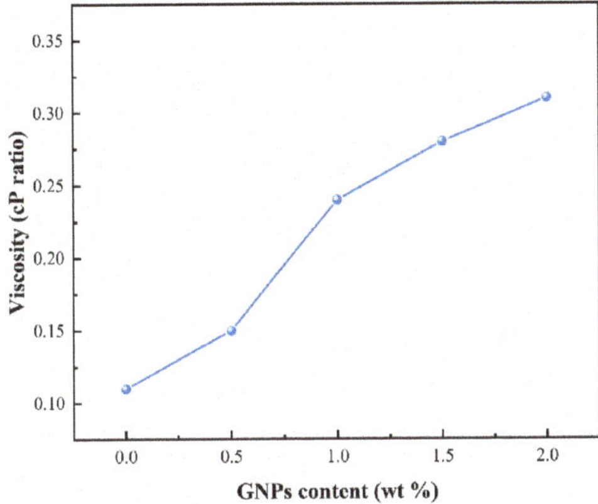

Figure 9. Viscosity of the composites with different contents of GNPs.

3.7. Photothermal Conversion Performance Analysis

To evaluate whether the PEG/GNPs composites are capable of absorbing and converting solar energy more effectively in comparison with pristine PEG when exposed to solar irradiation, we designed an experimental apparatus to conduct the photothermal conversion experiments on them, which was composed of two parts: photothermal conversion system and data acquisition system. The first system consisted of an insulated chamber, serial thermometer, temperature control device, temperature-sensing device, and a light source. The second system consisted of temperature logger with several thermocouples and a computer. Prior to testing, the sample melted at 85 °C was poured into a quartz beaker, then the thermocouple was put in the middle of the sample. During the process of cooling, it is imperative to keep the thermocouple in the correct position to ensure close contact with the sample. The variations in temperature of pristine PEG and PEG/GNPs-2% composite with the irradiation time are depicted in Figure 10. As can been observed, during the light irradiation period, the PEG/GNPs-2% composite can fully absorb and convert the light energy, as a result, the temperature exhibits a rapid rising trend. When the temperature achieves the melting value of the composite, a turning point appears in the temperature variation curve, indicating that the composite begins to process a phase change and store energy. After the illumination is removed, the temperature of the composite starts to decline, and the curve shows a constant temperature heat release platform, which indicates that the composite begins to crystallize. However, the temperature of pristine PEG rises slowly and barely reaches its melting point, exhibiting a low efficiency in utilizing the light energy. Therefore, compared with pure PEG, the PEG/GNPs nanocomposites show better performance in photothermal energy conversion.

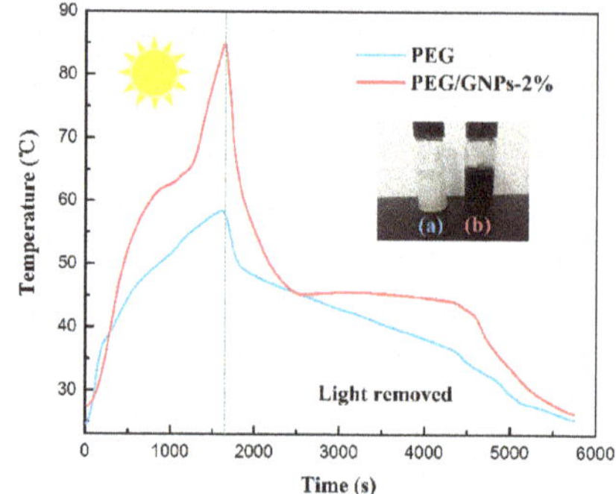

Figure 10. The variation in temperature with irradiation time of pure PEG and PEG/GNPs-2% composite.

To illustrate the reasons for the higher absorption efficiency of PEG/GNPs for solar irradiation compared with pure PEG, the Ultraviolet-visible-Near Infrared spectroscopy was used to measure the diffuse reflection spectra of PEG and PEG/GNPs composites. As shown in the Figure 11, the light absorption performance of PEG/GNPs composite is obviously much higher than that of pristine PEG in throughout at 200 to 2000 nm wavelength, suggesting that the PEG doped with GNPs make an outstanding increase in the solar light absorbance characteristic. Furthermore, the composite shows one total absorbance property in the different spectral region, which is essential to utilize the solar thermal energy to the greatest extent. This possibly can be attributed to a zero-bandgap structure of GNPs, which can theoretically absorb any wavelength, in addition, when the intensity of the incident light exceeds a certain threshold, the absorption of GNPs will reach saturation. As can be seen from photograph (inset in Figure 10), compared with PEG/GNPs composite, the pristine PEG has a white surface which can form a reflective layer that hinders the absorption of solar light. Accordingly, compared with pure PEG, the PEG/GNPs composites exhibit higher conductivity and better solar radiation absorption capacity, resulting in the more excellent performance in photothermal energy conversion. Therefore, the PEG/GNPs composites are expected to have great potential for solar energy conversion with respect to their better thermal conductivity and photothermal performance.

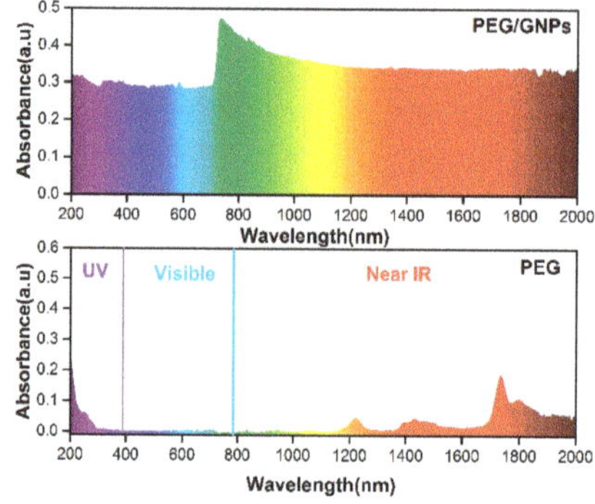

Figure 11. The Ultraviolet-visible-Near Infrared diffuse reflection spectra of pure PEG and PEG/GNPs-2% composite.

4. Conclusions

The PEG/GNPs nanocomposites with various contents of GNPs have been successfully prepared with the objective of investigating the enhanced thermal conductivity and photothermal performance of this nanocomposites. Samples of the PEG/GNPs nanocomposites were synthesized using a temperature-assisted solution blending method. SEM reveal that the GNPs are uniformly dispersed in the PEG matrix and construct a heat conduction way. FT-IR and XRD analysis show that no chemical interaction between PEG and GNPs happen, in addition, the suitable content of GNPs can promote the crystallization of PEG. The DSC results show that the phase change temperature and enthalpy are slightly influenced by GNPs content. In addition, the thermal conductivity of the composites is remarkably enhanced by the addition of GNPs, and adding 2 wt. % GNPs results in 146% increase in thermal conductivity and only 6.3% reduction in the latent heat of the PEG. The photothermal conversion performance of the PEG/GNPs nanocomposites is higher than that of pure PEG, which is ascribed to the nanocomposites possess better thermal conductivity and excellent visible light absorptivity in comparison with the pure PEG. The enhanced thermal conductivity property and photothermal conversion performance make the PEG/GNPs nanocomposites very promising for the application of solar energy conversion and storage.

Author Contributions: Data curation, H.W. and Y.G.; Formal analysis, X.L.; Investigation, H.W.; Methodology, H.W. and H.Z.; Resources, X.M.; Software, H.W.; Supervision, L.H.; Writing—original draft, H.W.; Writing—review & editing, H.W.

Funding: This work was jointly supported by National Natural Science Foundation of China (Grant Nos. 51508063, 51506018), the Science and Technology Research Program of Chongqing Municipal Education Commission (Grant No. KJ1705145).

Acknowledgments: The authors would like to express gratitude to thank the reviewers and editor for kindly giving revising suggestions.

References

1. Karaipekli, A.; Biçer, A.; Sarı, A.; Tyagi, V.V. Thermal characteristics of expanded perlite/paraffin composite phase change material with enhanced thermal conductivity using carbon nanotubes. *Energy Convers. Manag.* **2017**, *134*, 373–381. [CrossRef]

2. Barlev, D.; Vidu, R.; Stroeve, P. Innovation in concentrated solar power. *Sol. Energy Mater. Sol. Cells* **2011**, *95*, 2703–2725. [CrossRef]

3. Granqvist, C.G. Transparent conductors as solar energy materials: A panoramic review. *Sol. Energy Mater. Sol. Cells* **2007**, *91*, 1529–1598. [CrossRef]

4. Crabtree, G.W.; Lewis, N.S. Solar energy conversion. *Dalton Trans.* **2009**, *60*, 37–42.

5. Zeng, J.-L.; Zheng, S.-H.; Yu, S.-B.; Zhu, F.-R.; Gan, J.; Zhu, L.; Xiao, Z.-L.; Zhu, X.-Y.; Zhu, Z.; Sun, L.-X.; et al. Preparation and thermal properties of palmitic acid/polyaniline/exfoliated graphite nanoplatelets form-stable phase change materials. *Appl. Energy* **2014**, *115*, 603–609. [CrossRef]

6. Khan, Z.; Khan, Z.; Ghafoor, A. A review of performance enhancement of PCM based latent heat storage system within the context of materials, thermal stability and compatibility. *Energy Convers. Manag.* **2016**, *115*, 132–158. [CrossRef]

7. Qi, G.-Q.; Liang, C.-L.; Bao, R.-Y.; Liu, Z.-Y.; Yang, W.; Xie, B.-H.; Yang, M.-B. Polyethylene glycol based shape-stabilized phase change material for thermal energy storage with ultra-low content of graphene oxide. *Sol. Energy Mater. Sol. Cells* **2014**, *123*, 171–177. [CrossRef]

8. Fang, G.; Li, H.; Chen, Z.; Liu, X. Preparation and properties of palmitic acid/SiO_2 composites with flame retardant as thermal energy storage materials. *Sol. Energy Mater. Sol. Cells* **2011**, *95*, 1875–1881. [CrossRef]

9. Agyenim, F.; Hewitt, N.; Eames, P.; Smyth, M. A review of materials, heat transfer and phase change problem formulation for latent heat thermal energy storage systems (LHTESS). *Renew. Sustain. Energy Rev.* **2010**, *14*, 615–628. [CrossRef]

10. Kaizawa, A.; Kamano, H.; Kawai, A.; Jozuka, T.; Senda, T.; Maruoka, N.; Akiyama, T. Thermal and flow behaviors in heat transportation container using phase change material. *Energy Convers. Manag.* **2008**, *49*, 698–706. [CrossRef]

11. Yang, J.; Qi, G.-Q.; Liu, Y.; Bao, R.-Y.; Liu, Z.-Y.; Yang, W.; Xie, B.-H.; Yang, M.-B. Hybrid graphene aerogels/phase change material composites: Thermal conductivity, shape-stabilization and light-to-thermal energy storage. *Carbon* **2016**, *100*, 693–702. [CrossRef]

12. Amin, M.; Putra, N.; Kosasih, E.A.; Prawiro, E.; Luanto, R.A.; Mahlia, T.M.I. Thermal properties of beeswax/graphene phase change material as energy storage for building applications. *Appl. Therm. Eng.* **2017**, *112*, 273–280. [CrossRef]

13. Wu, W.F.; Liu, N.; Cheng, W.L.; Liu, Y. Study on the effect of shape-stabilized phase change materials on spacecraft thermal control in extreme thermal environment. *Energy Convers. Manag.* **2013**, *69*, 174–180. [CrossRef]

14. He, L.; Li, J.; Zhou, C.; Zhu, H.; Cao, X.; Tang, B. Phase change characteristics of shape-stabilized PEG/SiO$_2$ composites using calcium chloride-assisted and temperature-assisted sol gel methods. *Sol. Energy* **2014**, *103*, 448–455. [CrossRef]

15. Sarı, A.; Alkan, C.; Biçer, A. Synthesis and thermal properties of polystyrene-graft-PEG copolymers as new kinds of solid–solid phase change materials for thermal energy storage. *Mater. Chem. Phys.* **2012**, *133*, 87–94. [CrossRef]

16. Alkan, C.; Günther, E.; Hiebler, S.; Himpel, M. Complexing blends of polyacrylic acid-polyethylene glycol and poly(ethylene-co-acrylic acid)-polyethylene glycol as shape stabilized phase change materials. *Energy Convers. Manag.* **2012**, *64*, 364–370. [CrossRef]

17. Onder, E.; Sarier, N.; Ukuser, G.; Ozturk, M.; Arat, R. Ultrasound assisted solvent free intercalation of montmorillonite with PEG1000: A new type of organoclay with improved thermal properties. *Thermochim. Acta* **2013**, *566*, 24–35. [CrossRef]

18. Tang, Y.; Jia, Y.; Alva, G.; Huang, X.; Fang, G. Synthesis, characterization and properties of palmitic acid/high density polyethylene/graphene nanoplatelets composites as form-stable phase change materials. *Sol. Energy Mater. Sol. Cells* **2016**, *155*, 421–429. [CrossRef]

19. Deng, Y.; Li, J.; Qian, T.; Guan, W.; Li, Y.; Yin, X. Thermal conductivity enhancement of polyethylene glycol/expanded vermiculite shape-stabilized composite phase change materials with silver nanowire for thermal energy storage. *Chem. Eng. J.* **2016**, *295*, 427–435. [CrossRef]

20. Zhang, X.; Huang, Z.; Ma, B.; Wen, R.; Zhang, M.; Huang, Y.; Fang, M.; Liu, Y.; Wu, X. Polyethylene glycol/Cu/SiO$_2$ form stable composite phase change materials: Preparation, characterization, and thermal conductivity enhancement. *RSC Adv.* **2016**, *6*, 58740–58748. [CrossRef]

21. Harikrishnan, S.; Magesh, S.; Kalaiselvam, S. Preparation and thermal energy storage behaviour of stearic acid–TiO$_2$ nanofluids as a phase change material for solar heating systems. *Thermochim. Acta* **2013**, *565*, 137–145. [CrossRef]

22. Cataldi, P.; Athanassiou, A.; Bayer, I. Graphene Nanoplatelets-Based Advanced Materials and Recent Progress in Sustainable Applications. *Appl. Sci.* **2018**, *8*, 1438. [CrossRef]

23. Silakhori, M.; Fauzi, H.; Mahmoudian, M.R.; Metselaar, H.S.C.; Mahlia, T.M.I.; Khanlou, H.M. Preparation and thermal properties of form-stable phase change materials composed of palmitic acid/polypyrrole/graphene nanoplatelets. *Energy Build.* **2015**, *99*, 189–195. [CrossRef]

24. Hu, K.; Kulkarni, D.D.; Choi, I.; Tsukruk, V.V. Graphene-polymer nanocomposites for structural and functional applications. *Prog. Polym. Sci.* **2014**, *39*, 1934–1972. [CrossRef]

25. Pu, M.; Chen, P.; Wang, Y.; Zhao, Z.; Wang, C.; Huang, C.; Hu, C.; Luo, X. Strong enhancement of light absorption and highly directive thermal emission in graphene. *Opt. Express* **2013**, *21*, 11618–11627. [CrossRef]

26. Novoselov, K.S.; Geim, A.K.; Morozov, S.V.; Jiang, D.; Katsnelson, M.I.; Grigorieva, I.V.; Dubonos, S.V.; Firsov, A.A. Two-dimensional gas of massless Dirac fermions in graphene. *Nature* **2005**, *438*, 197. [CrossRef]

27. Nilsson, J.; Neto, A.H.; Guinea, F.; Peres, N.M. Electronic properties of graphene multilayers. *Phys. Rev. Lett.* **2006**, *97*, 266801. [CrossRef]

28. Nair, R.R.; Blake, P.; Grigorenko, A.N.; Novoselov, K.S.; Booth, T.J.; Stauber, T.; Peres, N.M.; Geim, A.K. Fine structure constant defines visual transparency of graphene. *Science* **2008**, *320*, 1308. [CrossRef]

29. Mehrali, M.; Latibari, S.T.; Mehrali, M.; Mahlia, T.M.I.; Metselaar, H.S.C.; Naghavi, M.S.; Sadeghinezhad, E.; Akhiani, A.R. Preparation and characterization of palmitic acid/graphene nanoplatelets composite with remarkable thermal conductivity as a novel shape-stabilized phase change material. *Appl. Therm. Eng.* **2013**, *61*, 633–640. [CrossRef]

30. Fang, X.; Fan, L.W.; Ding, Q.; Wang, X.; Yao, X.L.; Hou, J.F.; Yu, Z.T.; Cheng, G.H.; Hu, Y.C.; Cen, K.F. Increased Thermal Conductivity of Eicosane-Based Composite Phase Change Materials in the Presence of Graphene Nanoplatelets. *Energy Fuels* **2013**, *27*, 4041–4047. [CrossRef]

31. Yavari, F.; Fard, H.R.; Pashayi, K.; Rafiee, M.A.; Zamiri, A.; Yu, Z.; Ozisik, R.; Borcatasciuc, T.; Koratkar, N. Enhanced Thermal Conductivity in a Nanostructured Phase Change Composite due to Low Concentration Graphene Additives. *J. Phys. Chem. C* **2011**, *115*, 8753–8758. [CrossRef]

32. Li, J.F.; Lu, W.; Zeng, Y.B.; Luo, Z.P. Simultaneous enhancement of latent heat and thermal conductivity of docosane-based phase change material in the presence of spongy graphene. *Sol. Energy Mater. Sol. Cells* **2014**, *128*, 48–51. [CrossRef]

33. Su, Q.; Pang, S.; Alijani, V.; Li, C.; Feng, X.; Müllen, K. Composites of Graphene with Large Aromatic Molecules. *Adv. Mater.* **2009**, *21*, 3191–3195. [CrossRef]

34. He, L.; Wang, H.; Yang, F.; Zhu, H. Preparation and properties of polyethylene glycol/unsaturated polyester resin/graphene nanoplates composites as form-stable phase change materials. *Thermochim. Acta* **2018**, *665*, 43–52. [CrossRef]

35. Yu, A.; Ramesh, P.; Itkis, M.E.; Elena Bekyarova, A.; Haddon, R.C. Graphite Nanoplatelet—Epoxy Composite Thermal Interface Materials. *J. Phys. Chem. C* **2007**, *111*, 7565–7569. [CrossRef]

36. Yu, Z.T.; Fang, X.; Fan, L.W.; Wang, X.; Xiao, Y.Q.; Zeng, Y.; Xu, X.; Hu, Y.C.; Cen, K.F. Increased thermal conductivity of liquid paraffin-based suspensions in the presence of carbon nano-additives of various sizes and shapes. *Carbon* **2013**, *53*, 277–285. [CrossRef]

37. He, Q.; Wanga, S.; Liu, Y. Experimental study on thermophysical properties of nanofluids as phase-change material (PCM) in low temperature cool storage. *Energy Convers. Manag.* **2012**, *64*, 199–205. [CrossRef]

38. Lin, C.-Y.; Wang, J.-C.; Chen, T.-C. Analysis of suspension and heat transfer characteristics of AlO nanofluids prepared through ultrasonic vibration. *Appl. Energy* **2011**, *88*, 4527–4533. [CrossRef]

Article

Experimental Investigation of Freezing and Melting Characteristics of Graphene-Based Phase Change Nanocomposite for Cold Thermal Energy Storage Applications

Shaji Sidney [1], Mohan Lal Dhasan [1], Selvam C. [2] and Sivasankaran Harish [3,*

[1]   Department of Mechanical Engineering, Anna University, College of Engineering Campus, Chennai 600 025, India; shajisidney@gmail.com (S.S.); mohanlal@annauniv.edu (M.L.D.)

[2]   Department of Mechanical Engineering, SRM Institute of Science and Technology, Kattankulathur, Chennai 603 203, India; selvammech87@gmail.com

[3]   International Institute for Carbon-Neutral Energy Research, Kyushu University, Nishi-ku, Fukuoka 819-0395, Japan

*   Correspondence: harish@i2cner.kyushu-u.ac.jp; Tel.: +81-92-802-6730

Received: 15 January 2019; Accepted: 28 February 2019; Published: 15 March 2019

Abstract: In the present work, the freezing and melting characteristics of water seeded with chemically functionalized graphene nanoplatelets in a vertical cylindrical capsule were experimentally studied. The volume percentage of functionalized graphene nanoplatelets varied from 0.1% to 0.5% with an interval of 0.1%. The stability of the synthesized samples was measured using zeta potential analyzer. The thermal conductivity of the nanocomposite samples was experimentally measured using the transient hot wire method. A ~24% (maximum) increase in the thermal conductivity was observed for the 0.5% volume percentage in the liquid state, while a ~53% enhancement was observed in the solid state. The freezing and melting behavior of water dispersed with graphene nanoplatelets was assessed using a cylindrical stainless steel capsule in a constant temperature bath. The bath temperatures considered for studying the freezing characteristics were $-6\,^{\circ}$C and $-10\,^{\circ}$C, while to study the melting characteristics the bath temperature was set as 31 $^{\circ}$C and 36 $^{\circ}$C. The freezing and melting time decreased for all the test conditions when the volume percentage of GnP increased. The freezing rate was enhanced by ~43% and ~32% for the bath temperatures of $-6\,^{\circ}$C and $-10\,^{\circ}$C, respectively, at 0.5 vol % of graphene loading. The melting rate was enhanced by ~42% and ~63% for the bath temperatures of 31 $^{\circ}$C and 36 $^{\circ}$C, respectively, at 0.5 vol % of graphene loading.

Keywords: nanocomposite; melting; freezing; graphene; thermal conductivity

---

1. Introduction

The world is facing a lot of challenges related to storing and retrieving energy and fulfilling the pressing energy demands. Heat is the main form of energy which can be stored, and this is achieved in the form of latent heat using phase change materials (PCM). The oldest form of thermal energy storage (TES) probably involves harvesting ice from lakes and rivers and storing it in well-insulated warehouses throughout the year for use in almost all tasks that mechanical refrigeration satisfies today, including food preservation, cooling of drinks, and air-conditioning. A variety of TES techniques have been developed over the past decades. Today compressed-air storage and batteries are mostly used to meet many of the thermal energy storage requirements.

Instead of storing electrical energy in a battery or as compressed air, thermal energy storage using water-based ice is one of the most ancient modes of energy storage and is considered to be the most efficient and economic mode of energy storage, as it eliminates the recurring expenses incurred for the replacement of batteries. Water can be used as an effective thermal energy storage material due to its higher thermal conductivity and excellent freezing/melting characteristics. Cold energy stored in ice can be effectively used to remove heat from another fluid in a secondary circuit. Water acts as a good thermal energy storage material in various industries such as the dairy industry for chilling milk [1,2], and the pharmaceutical [3] and chemical industries for transportation and storage without having to depend on batteries.

The refrigeration sector has now evolved the use of DC powered compressors that directly utilize the use of solar energy eliminating inverters. Likewise, few researches have started using DC powered compressors without batteries autonomously depending on ice-based thermal storage [1–4]. Pedersen and Katic (2016) confirmed that the energy content in ice produced by the DC compressor was higher than the energy content in a lead-acid battery, in terms of both volume and weight.

However, further research has been going on as regards replacing water with other fluids or choosing the best additives so as to improve the freezing and melting characteristics enabling one to store more thermal energy and to have smaller thermal storage devices. One of the most suitable methods is to add highly thermal conductive material to the water. Among the materials used, metal and metal oxides in nano-metric sizes exhibit excellent thermal transport properties. Owing to the higher density of metal and metal oxide powders, carbon-based nanomaterials are widely used because of their high aspect ratio. Hence, adding carbon-based nanomaterials is an effective way of increasing the thermal energy storage of water. Thus, this study is focused on the experimental investigation into the freezing and melting characteristics of graphene-based water for thermal energy storage applications such as in milk chilling and the chemical industries.

Guruprasad et al. (2017) [5] suggested that for medium and low temperature systems, the use of phase change materials (PCM) can be cost effective and will improve the thermal conductivity of thermal energy storage materials and play a major role in increasing the charging and discharging rate. They inferred that the thermal enhancement achieved with carbon-based nanostructures is better than those with metallic and metal oxide. The maximum enhancement in thermal conductivity obtained by Sathish Kumar et al. (2016) [6] was 9.5% for 0.6 wt. % of graphene nanoplatelets dispersed in water with the use of surfactants. A 24% reduction in the solidification time was observed for the nanocomposite with 0.6 wt. % of GnP. The experiments conducted by Ahammed et al. (2016) [7] showed an increase of 5.23% in thermal conductivity of graphene–water nanocomposite, prepared using surfactant, when the volume concentration of nanoparticles is changed from 0.05% to 0.1%, and a 14.56% enhancement was observed when the volume concentration increased by three times. Harikrishnan et al. (2014) [8] inferred that the latent heat of composite PCMs is lower than that of base material for both melting and freezing and the maximum changes are 3.56% and 3.82%, respectively. The thermal conductivity of graphene–water nanocomposite is found to be higher when compared with that of the metal oxide nanoparticles and is lower when compared with that of pure metallic nanoparticles. However, the use of pure metallic nanoparticles in fluids causes stability problems. Hence, Ahammed et al. (2016) [9] suggested that instead of using a high-volume concentration of metal oxide and pure metal nanoparticles, a low-volume concentration of graphene can be used as the heat transfer fluid to enhance thermal conductivity. Harish et al. (2015) [10] treated graphene nanoparticles with concentrated nitric acid to avoid the use of surfactants. A maximum thermal conductivity enhancement of ~230% was measured in lauric acid treated with the acid graphene nanoinclusions for 1 vol %.

As a result of the ever-growing demand for energy, there is a need for energy storage in PCMs. The PCMs are usually encapsulated in containers/capsules. Different researches have used different geometrical shapes for the capsules such as cylinders, spheres, pyramids, cones, rectangles, and cuboids with different materials like stainless steel, aluminum, copper, polypropylene, and polyolefin for numerical and experimental studies. The material selection was based on the property of the PCM used and the potential applications [11–16]. Yoon et al. (2001) [17] studied the freezing properties of water in a circular cylinder kept horizontally. During the initial phase of freezing, an annular ice layer started growing on the surface of the cylinder at a higher rate. This was followed by the asymmetric ice layer at a medium cooling rate and finally an instantaneous ice layer growing over the whole region at a low cooling rate. Kalaiselvam et al. (2008) [18] performed an analytical analysis in the freezing and melting process of different PCMs encapsulated in a cylindrical capsule. The presence of heat generation enhanced the freezing time whilst also hastening the melting time. Total freezing time was subject to Stefan's Number and heat generation parameters, whereas complete melting time depended on equivalent thermal conductivity.

Nanotechnology is being used in many applications to provide more efficient energy transfer. The application of nanocomposites in heat exchanging devices appears promising with these characteristics. In this context, the use of nanoparticles in water provides a scope for performance improvement in thermal storage for an ice bank tank. The main objective of this work is to study the freezing and melting characteristics of graphene–water nanocomposite in a vertical cylindrical capsule and compare it with base fluid. Graphene has been widely used in many applications since its discovery by Novoselov et al. [19] because of its structure i.e., a single-atom-thick sheet of hexagonally arrayed sp^2-bonded carbon atoms. Graphene possesses remarkable thermophysical properties due to its large specific surface area (50–750 m^2/g) and extremely high thermal conductivity (3000–5000 W/m K) [20–28]. The thermophysical properties of graphene nanocomposite, such as thermal conductivity, are also stable for the temperatures ranging from $-10\ °C$ to $40\ °C$, the zeta potential being a reason for this.

2. Materials and Methods

The thermal conductivity of the functionalized graphene–water nanocomposites were measured experimentally. The study was carried out using chemically treated graphene nanoplatelets to avoid the use of surfactants which were used to improve stability of the nanocomposites.

2.1. Preparation of Graphene Nanocomposite

The nanocomposite was prepared prior to the experimental work using the two-step method. The essential requirements for nanocomposites are: a stable suspension, adequate durability, negligible agglomeration of particulates, no chemical change in the particulates or fluid, etc. The required quantity of graphene nanoplatelets was purchased from XG Sciences (Lansing, MI, USA). The scanning electron microscopy (SEM) (FEI 3D Versa Dual Beam, Hillsboro, Oregon, USA) and transmission electron microscopy (TEM) (JEOL JEM-2000EX, Akishima, Tokyo, Japan) visualization of the GnP used is shown in Figure 1. The GnP–H_2O was prepared using the covalent functionalization method. The GnPs were chemically functionalized with concentrated nitric acid (68 wt. %) to improve the dispersion of the particles and to avoid the use of surfactants. Five grams of graphene nanoplatelets were dispersed in 250 mL of concentrated nitric acid taken in a conical flask and then refluxed at a temperature of 100 °C for 2 h. For uniform dispersion, the fluid was stirred using a magnetic stirrer. To maintain a constant temperature during the process, the conical flask was placed in a constant temperature bath, which was maintained at 100 °C. After 2 h, the conical flask was taken out from the oil bath and allowed to cool down slowly to room temperature. The nanoplatelets were filtered then washed with DI water and dried in a furnace at 160 °C [10]. Nitric acid treatment was used to chemically modify the surface of graphene platelets in order to increase the surface-active sites for electrochemical reactions because of the hydrophobic nature of GnP. The nitric acid treatment introduced more oxygen/nitrogen-containing functional groups onto the graphene surface, and clearly enhanced the hydrophilicity of the graphene. This was to promote the wettability of graphene when it was dispersed in DI water, which was the base fluid. For the experimental work, the volume percentages of GnP were 0.1%, 0.2%, 0.3%, 0.4%, and 0.5%. Depending on the volume percentage, the required quantity of chemically functionalized GnP was added to the DI water and was stirred for 30 min using a magnetic stirrer. After stirring, the fluid was ultra-sonicated using a digital sonicator (Qsonica, Newtown, CT, USA) for 2 h to enhance the stability. The graphene nanocomposites, thus prepared, were kept for observation and no particle sedimentation was observed at the bottom of the bottle even after two weeks.

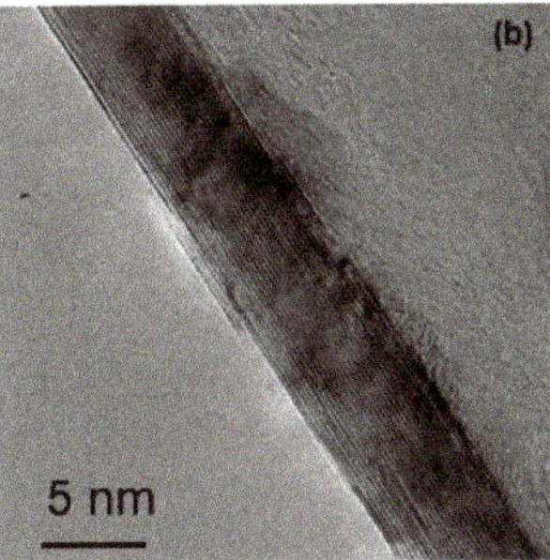

Figure 1. (**a**) SEM visualization of GnP; (**b**) TEM visualization of GnP.

2.2. Stability Analysis

A zeta potential measurement was carried out in order to ensure the stability of the prepared nanocomposite. The general reference for average zeta potential values are considered to be more negative than −30 mV or more positive than +30 mV in order to predict the stability of dispersion, while poor stability will show a value below 20 mV [28,29]. Figure 2 shows that the stability of nanocomposite (0.5 vol. %) lies in the excellent stability region with a zeta potential peak of −69.4 mV. The nanoparticles are highly electronegative, and this indicates the excellent stability of the nanocomposite after acid treatment of GnP. The stability of the nanocomposite (0.5 vol. %) was measured after several cycles (>10) after freezing and melting experimentations.

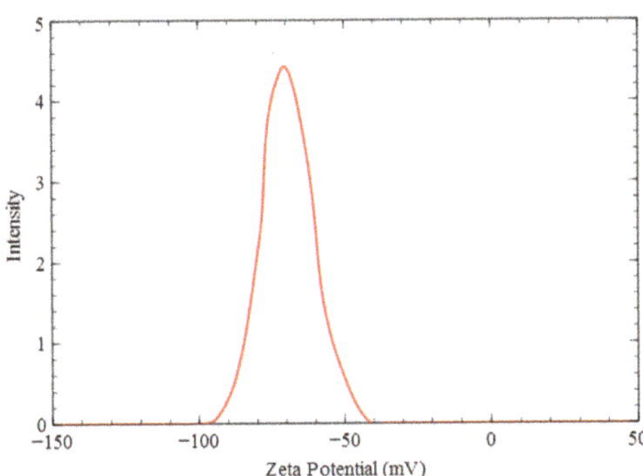

Figure 2. Zeta potential distribution of the nanocomposite at 0.5 vol. %.

2.3. Thermal Conductivity Measurement

The thermal conductivity of the nanocomposite was measured in the temperature range of −10 °C to 40 °C using the KD2 Pro thermal analyzer (Decagon devices, Pullman, WA, USA), which works on the principle of the transient hot wire method. The nanocomposite sample was poured in a small container and the KS1 sensor probe was inserted at the center of the container. The desired test temperatures of the samples were achieved by immersing the container in a refrigerated/heating circulator bath system, which maintains the temperature of the surrounding fluid with an accuracy of ±0.03 °C. The sensor used to measure the thermal conductivity in the KD2 Pro apparatus was the KS-1 sensor (60 mm long, 1.3 mm diameter) with an accuracy of ±10%. The sensor is integrated with a heating element and a thermo-resistor in the core and is connected to a microprocessor to control and conduct the measurements. The precise results were obtained by keeping the probe in the fluid sample continuously for 20 min, after attaining the desired equilibrium temperature. Five measurements were taken for each sample, to ensure the repeatability and accuracy of the result. While measuring the thermal conductivity in the solid phase of the nanocomposite, a thermal grease was applied on the surface of the KS-1 sensor as per the instructions by the KD2 Pro thermal analyzer manual. Holes were drilled in the solid nanocomposite then the KS-1 sensor with thermal grease was inserted to measure the thermal conductivity.

2.4. Experimental Test Facility for Freezing and Melting Characteristics

Figure 3 shows a schematic representation of the experimental test facility which was used to study the freezing/melting characteristics of the nanocomposite. The experiments consisted of charging and discharging of nanocomposite in a stainless steel (SS) cylindrical capsule. A cold bath was used for the freezing/charging experiments and a hot bath was used for the melting/discharging experiment. Experiments were carried out based on the test matrix as given in Section 2.5.

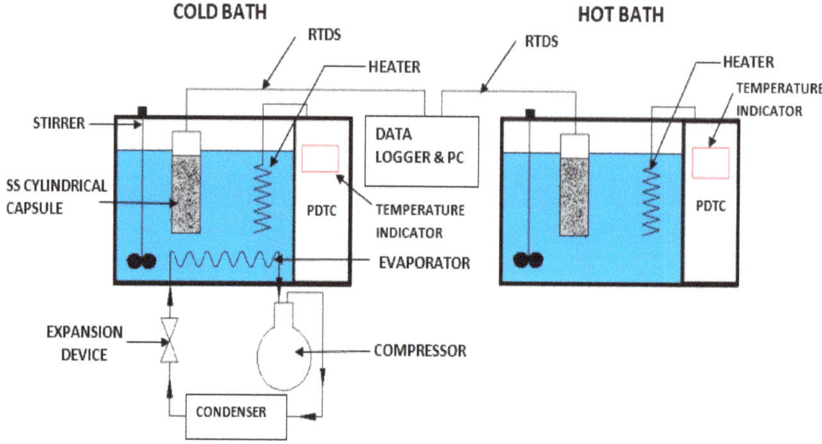

Figure 3. Experimental setup for the freezing/melting study.

A stainless steel cylindrical capsule was used to carry out the freezing and melting experiments as in most thermal storage systems. The capsule was made of SS because most food storage appliances are made of SS [1]. The vertical cylindrical capsule had an inner diameter of 46 mm and height of 120 mm of a total capacity of 200 mL with a wall thickness of 1.5 mm. The temperature sensor in the middle of the cylinder was considered for the freezing and melting study. The nanocomposite was poured inside the capsule and then it was placed in the freezing bath. The capsule was filled to only 80% of its full capacity i.e., 160 mL, to account for the volume of RTDs (PT100 type) and the expansion of water when it becomes ice. The experimental setup for freezing (charging) consisted of a freezing unit with an evaporator tank, a condensing unit, a proportional differential temperature controller (PDTC), a cylindrical capsule, a computer, and a data logger. The cold bath was filled with a mixture of water and ethylene glycol so that negative temperatures could be achieved without freezing the cold bath. The transient temperature variations of the nanocomposites were measured and recorded continuously every 30 s using a data logger. The experimental setup for melting (discharging) consisted of a hot bath with a storage tank, a heating coil, a PDTC, a cylindrical capsule, a computer, and a data logger.

2.5. Test Matrix and Working Procedure

Freezing and melting experiments were carried out by placing a cylindrical SS capsule filled with nanocomposite in a constant temperature bath. The bath temperatures used were −6 °C and −10 °C for freezing the nanocomposite to −3 °C, while 31 °C and 36 °C were considered for the melting experimentations until the nanocomposites reached 30 °C. Table 1 shows the test matrix for the freezing and melting experiments.

Table 1. Freezing and melting test matrix.

Volume Percentage of Functionalised GnP	Initial Temperature of the Sample	Bath Temperatures for Freezing	Sample Temperature after Freezing	Bath Temperature for Melting	Sample Temperature after Melting
0%, 0.1%, 0.2%, 0.3%, 0.4% & 0.5%	32 °C	−6 °C & −10 °C	−3 °C	31 °C & 36 °C	30 °C

During the charging process, the test sample was placed inside the cold bath tank where the temperature was maintained below the freezing temperature of the nanocomposite. In the beginning of the charging process, the nanocomposite was sensibly cooled until it reached the freezing temperature. At the freezing temperature, the latent heat was absorbed by the nanocomposite and it underwent phase change from liquid to solid. After freezing, the nanocomposite was again sensibly cooled until the bath temperature was achieved.

During the discharging process, the cylindrical capsule with the fully solidified nanocomposite was placed inside the storage tank where the temperature was maintained above the melting temperature of the nanocomposite. In the beginning of the discharging process, the nanocomposite was sensibly heated until it reached the melting temperature. At the melting temperature, the latent heat was released by the nanocomposite and it underwent phase change from solid to liquid. After melting, the nanocomposite was again sensibly heated until the bath temperature was achieved.

3. Results and Discussion

3.1. Thermal Conductivity

Initially, the thermal conductivity of pure DI water was experimentally measured at temperatures ranging from −10 to 40 °C. Then, the experimental data were compared with the standard data to validate the measurement practice [30–32]. It was found that the measured values matched-up with the standard values. The average percentage deviation of measured values from the standard values of the base fluid was ±2%.

Figure 4 shows the variation in thermal conductivity of the nanocomposite in a liquid state, and it is observed that the graphene nanocomposite has higher thermal conductivity when compared to the base fluid. The dispersion of GnP significantly improved the thermal conductivity of the nanocomposite. The reason behind this enhancement of thermal conductivity is the nano-size of the GnP and the two-dimensional geometry of the GnP that increases the exposure to the base fluid.

By increasing the volume percentage of the GnP, the thermal conductivity was found to increase. The average increase in thermal conductivity compared to base fluid were found to be 11.01%, 13.38%, 17.23%, 20.96%, and 23.95% respectively, for the considered volume percentages of the GnP in the liquid state. The thermal conductivity for the 0.5% volume percentage was 23.95% higher compared to that of the base fluid; whereas Selvam et al. (2016) obtained a thermal conductivity 16% higher than the base fluid at the same concentration. The superior thermal conductivity is due to the use of chemically functionalized GnP instead of dispersing surfactants in the base fluid for better stability. Whereas, in the solid state, as shown in Figure 3, the average surge in thermal conductivity compared to base fluid are 18.67%, 25.7%, 34.75%, 45.08%, and 53.05%, respectively, for the considered volume percentages of GnP.

Thermal conductivity in the solid state was higher than in the liquid state. The reason for this sudden increase in thermal conductivity when the fluid changes to a solid state is due to the orderly solid structure that causes accelerated molecular vibrations. The sudden fall in thermal conductivity when the solid state changes to liquid state might be caused by the orderly stable microstructure in the solid turning into a disorderly structure in the liquid state. Table 2 shows the tabulated comparison of thermal conductivity of the nanocomposite with respect to temperature and % volume fraction. The reason for thermal conductivity enhancement is attributed to the high aspect ratio, 2-D geometry, and stiffness of graphene.

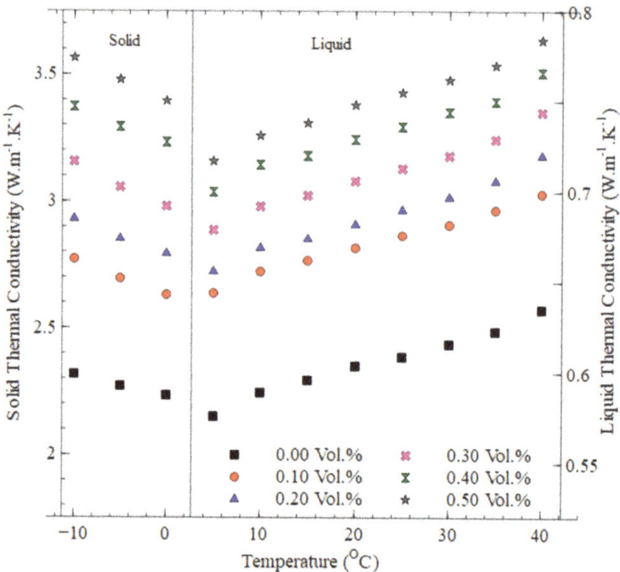

Figure 4. Variation of thermal conductivity with respect to temperature for different volume percentages.

The acid treatment of GnP plays a significant role in the enhancement of thermal conductivity in the nanocomposites. Instead of adding surfactant, nitric acid treatment was used to chemically modify the surface of GnP in order to increase the surface-active sites for electrochemical reactions and also to improve the stability. As surfactants not used, the effects of degradation in thermal conductivity while adding surfactant completely avoided in this nanocomposite.

Figure 4 shows the variation in thermal conductivity of the nanocomposite in liquid state and it is observed that the graphene nanocomposite has higher thermal conductivity when compared to the base fluid. The dispersion of GnP significantly improved the thermal conductivity of the nanocomposite. The reason behind this enhancement of thermal conductivity is the nano-size of the GnP and the two-dimensional geometry of GnP that increases the exposure to the base fluid.

By increasing the volume percentage of GnP, the thermal conductivity is found to increase. The average increase in thermal conductivity compared to base fluid are found to be 11.01%, 13.38%, 17.23%, 20.96% and 23.95% respectively for the considered volume percentage of GnP in the liquid state. The thermal conductivity for the 0.5% volume percentage is 23.95% higher compared to that of the base fluid, whereas Selvam et al. (2016) obtained a thermal conductivity 16% higher than the base fluid at the same concentration. The superior thermal conductivity is due to the use of chemically functionalized GnP instead of dispersing surfactants in the base fluid for better stability. Whereas, in the solid state as shown in Figure 4, the average surge in thermal conductivity compared to base fluid are 18.67%, 25.7%, 34.75%, 45.08% and 53.05% respectively for the considered volume percentage of GnP.

Table 2. Thermal conductivity of nanocomposite with respect to temperature and vol.%.

Temp (°C)	DI Water Standard K (W/mK)	DI Water Measured K (W/mK)	Nanocomposite (0.1%)		Nanocomposite (0.2%)		Nanocomposite (0.3%)		Nanocomposite (0.4%)		Nanocomposite (0.5%)	
			K (W/mK)	Percentage Increase (%)	K (W/mK)	Percentage Increase (%)	K (W/mK)	Percentage Increase (%)	K (W/mK)	Percentage Increase (%)	K (W/mK)	Percentage Increase (%)
−10	2.30	2.317	2.771	19.59	2.930	26.46	3.156	36.21	3.371	45.51	3.565	53.87
−5	2.25	2.27	2.694	18.67	2.851	25.59	3.055	34.58	3.292	45.02	3.479	53.24
0	2.22	2.233	2.63	17.77	2.792	25.05	2.980	33.46	3.232	44.73	3.395	52.04
5	0.57	0.576	0.644	11.80	0.656	13.87	0.679	17.8	0.700	21.53	0.717	24.42
10	0.58	0.589	0.656	11.37	0.669	13.57	0.692	17.52	0.715	21.34	0.731	24.19
15	0.589	0.596	0.662	11.07	0.674	13.03	0.698	17.1	0.720	20.83	0.738	23.85
20	0.598	0.604	0.669	10.76	0.682	12.95	0.706	16.81	0.729	20.71	0.748	23.79
25	0.607	0.609	0.676	11.00	0.690	13.33	0.713	17.05	0.736	20.85	0.755	23.94
30	0.615	0.616	0.682	10.71	0.697	13.2	0.720	16.89	0.744	20.7	0.762	23.78
35	0.623	0.623	0.690	10.75	0.706	13.3	0.729	16.95	0.750	20.45	0.770	23.54
40	0.63	0.635	0.699	10.07	0.720	13.45	0.744	17.15	0.766	20.62	0.784	23.46

Thermal conductivity in solid state was higher than in the liquid state. The reason for this sudden increase in thermal conductivity when fluid turns to solid state is due to the orderly solid structure that causes better accelerated molecular vibrations. The sudden fall in thermal conductivity when solid state changes to liquid state might be caused by the orderly stable microstructure in solid turning into a disorderly structure in liquid state. Table 2 shows the tabulated comparison of thermal conductivity of nanocomposite with respect to temperature and % volume fraction. The background for the thermal conductivity enhancement is an attribute to the high aspect ratio, 2-D geometry and stiffness of graphene.

The acid treatment of GnP plays a significant role in the enhancement of thermal conductivity in the nanocomposites. Instead of adding surfactant, Nitric acid treatment was used to chemically modify the surface of GnP in order to increase the surface-active sites for electrochemical reactions and also to improve the stability. As surfactants are not used, the effects of degradation in thermal conductivity while adding surfactant are completely avoided in this nanocomposite.

3.2. Freezing and Melting Characteristics

3.2.1. Freezing Process

Initially the nanocomposite, which is in the liquid phase at room temperature (32 °C), gets sensibly cooled to the freezing temperature by placing it inside the cold bath maintained at the desired temperature. After sensible cooling, the nanocomposite begins to solidify starting from the outermost surface, which is exposed to the heat-conducting surface. As a result, the outermost layer starts to solidify first. The solidification process continues until the midpoint of nanocomposite solidifies. After complete solidification, sensible cooling of the solidified sample takes place until the nanocomposite reaches −3 °C. The bath temperatures for freeing experiments were −6 °C and −10 °C.

Figure 5 shows the freezing curve at −6 °C and −10 °C for the base fluid and nanocomposites for different volume percentages of GnP from 32 °C to −3 °C. The time taken by the base fluid to solidify was 91.5 min, whereas the time taken by the 0.1%, 0.2%, 0.3%, 0.4%, and 0.5% vol. fraction nanocomposites to solidify was 77 min, 71 min, 65 min, 59 min, and 55 min, respectively. Thus, the addition of nanoparticles aided to reduce the freezing time by 15.30%, 22.40%, 28.96%, 35.52%, and 39.89% for the 0.1%, 0.2%, 0.3%, 0.4% and 0.5% vol. fractions, respectively.

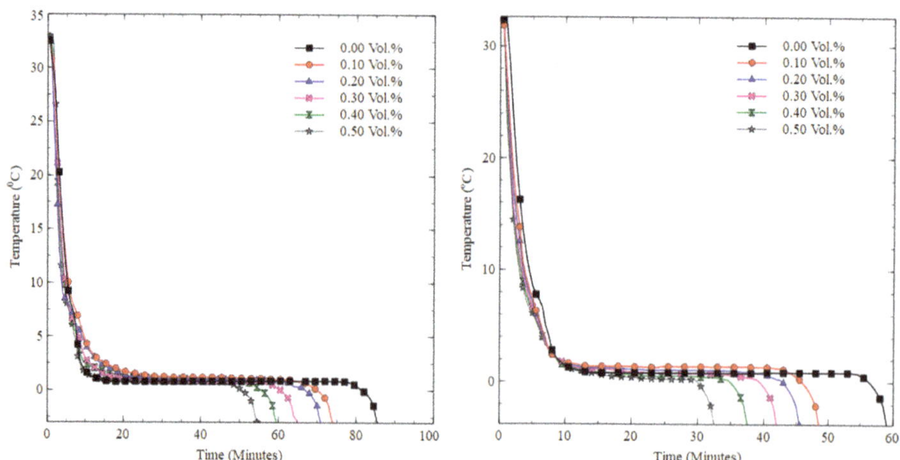

Figure 5. Freezing curves at from 32 °C to −3 °C with a bath temperature of −6 °C and −10 °C.

Similarly, at the bath temperature of −10 °C, the time taken to solidify the base fluid was 55.5 min, whereas for the 0.1%, 0.2%, 0.3%, 0.4%, and 0.5% vol. fractions of nanocomposites, the time taken was

46 min, 43.5 min, 40 min, 36 min, and 31 min, respectively. Thus, the addition of nanoparticles reduced the freezing time by 12.25%, 15.48%, 16.12%, 25.16%, and 31.61% for the 0.1%, 0.2%, 0.3%, 0.4%, and 0.5% vol. fractions, respectively.

3.2.2. Melting Process

The solidified nanocomposite is sensibly heated up to the melting temperature when placed inside the hot bath. After sensible heating, the nanocomposite begins to melt starting from the outermost surface, which is exposed to the heat-conducting surface. Consequently, the outermost layer starts to melt first. The melting process continues until the midpoint of the nanocomposite melts completely. Then the sensible heating of the liquid sample takes place until it is in thermal equilibrium with the hot bath temperature. The melting experiment is also carried out for the base fluid as well as the nanocomposites. The bath temperatures for the melting experiments were 31 °C and 36 °C, respectively. The melting process was carried out until the nanocomposite sample reached 30 °C.

Figure 6 shows the comparison of melting curves from −3 °C to 30 °C for the base fluid and nanocomposites, kept at bath temperatures of 31 °C and 36 °C. The sensible heating process was faster compared to the latent process. The time taken by the DI water to melt completely was 6 min, whereas the time taken by the 0.1%, 0.2%, 0.3%, 0.4%, and 0.5% vol. fractions of nanocomposites was 5.5 min, 5 min, 4.5 min, 4 min, and 3.5 min, respectively. Thus, the addition of the nanoparticles reduced the melting time by 8.33%, 17.67%, 25%, 33.33%, and 41.67% for the 0.1%, 0.2%, 0.3%, 0.4%, and 0.5% vol. fractions, respectively. Similar trends were observed when the samples were melted from −3 to 30 °C when kept in a hot bath at 36 °C. Under this condition, the time taken by DI water to melt completely was 4 min, whereas for the 0.1%, 0.2%, 0.3%, 0.4%, and 0.5% vol. fractions of nanocomposites, the time taken was 3.5 min, 3 min, 2.5 min, 2 min, and 1.5 min, respectively. Thus, the addition of nanoparticles reduced the melting time by 12.5%, 25%, 37.5%, 50%, and 62.5% for the 0.1%, 0.2%, 0.3%, 0.4%, and 0.5% vol. fractions, respectively.

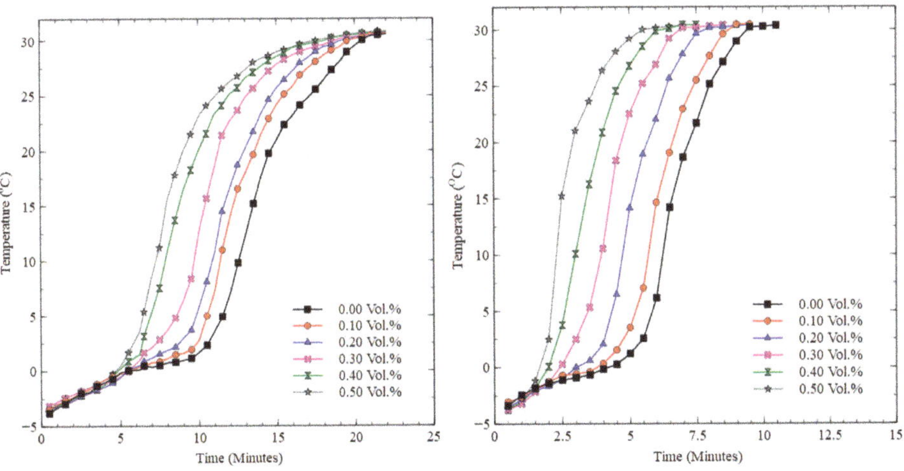

Figure 6. Melting curves from −3 °C to 30 °C with a bath temperature at 31 °C and 36 °C.

During melting, the solidified fluid near the walls absorbs heat and starts to melt. Conduction is the dominant mechanism in the initial time when the thickness of the liquid layer is so thin. The thickness of the liquid layer increases and buoyancy force is developed relative to time. The difference between the solid and liquid density causes the melted water to sink towards the bottom of the capsule and to consequently push the ice up. It intensifies the convection force and accelerates the melting rate. However, it can be noticed that the thermal conductivity of the liquid is lower than the solid, and therefore the heat conduction in the liquid is lower than in the solid. Thus, as the thickness of the liquid layer increases, the heat transfer conduction is reduced, and conversely the natural convection is enhanced. Therefore, melting is accelerated because of this natural convection in the liquid [33].

During freezing, the liquid near the walls rejects heat to the surrounding cold bath and starts to solidify. As time passes, the thickness of the solid layer increases. Water is denser than ice, which causes ice to float. Therefore, water gets accumulated at the bottom of the capsule. It can be seen that in the initial period of time, the solidification rate is high and conduction is the dominant mechanism of heat transfer between liquid and cold surfaces. As time progresses, a greater amount of liquid becomes solid and therefore the solid layer near the cold surfaces becomes thicker. Although the thermal conductivity of the solid is higher than the liquid, the solid layer imposes thermal resistance for heat conduction from the cold surface to the warm liquid. Thus, heat conduction decreases by increasing the thickness of the solid layer. The solidification rate reduces gradually during the process, especially near the end of process where the solid layer covers the whole of the capsule except a small region [33]. In both the freezing and melting experimentations, it was observed that the addition of nanoparticles enhanced the freezing and melting rates, and the maximum enhancement was observed with the 0.5% volume percentage of GnP.

4. Conclusions

Water-based graphene nanocomposites were chemically prepared using the covalent functionalization method and thermal conductivity was measured experimentally. The freezing and melting characteristics of the prepared nanocomposites were studied experimentally by varying the bath temperatures and volume percentages of the GnP. The results showed that the addition of GnP nanoplatelets increased the thermal conductivity of all volume fractions; a maximum enhancement of 23.95% was observed for the 0.5% volume fraction in the liquid state. Similarly, in the solid state, the maximum thermal conductivity enhancement was 53.05% for the 0.5% volume fraction. The freezing and melting time decreased for all the test conditions when the volume percentage of GnP increased.

Author Contributions: Conceptualization S.H and M.L.D.; Methodology, S.S. and S.C.; Formal analysis, S.S. and S.C. and S.H; writing—original draft preparation, S.S.; writing—review and editing, S.H., S.C. and M.L.D.; supervision M.L.D.

Funding: S.S. and M.L.D. would like to acknowledge the University Grants Commission (UGC), Government of India for providing the financial support to the first author under the Maulana Azad National Fellowship Scheme (F1-17.1/2016-17/MANF-2015-17-TAM-51497). S.H acknowledges the support of International Institute for Carbon-Neutral Energy Research (WPI-I²CNER).

Conflicts of Interest: The authors declare no conflict of interest.

References

1. de Blas, M.; Appelbaum, J.; Torres, J.L.; García, A.; Prieto, E.; Illanes, R. A Refrigeration Facility for Milk Cooling Powered by Photovoltaic Solar Energy. *Prog. Photovolt. Res. Appl.* **2003**, *11*, 467–479. [CrossRef]
2. Torres-Toledo, V.; Meissner, K.; Coronas, A.; Müller, J. Performance characterisation of a small milk cooling system with ice storage for PV applications. *Int. J. Refrig.* **2015**, *60*, 81–91. [CrossRef]
3. Pedersen, P.H.; Maté, J. SolarChill Vaccine Cooler and Refrigerator: A Breakthrough Technology. *Industria Informatione. Refrig. Air Cond.* **2006**, *300* (Suppl. 1), 17–19.
4. Axaopoulos, P.J.; Theodoridis, M.P. Design and experimental performance of a PV Ice-maker without battery. *Sol. Energy* **2009**, *83*, 1360–1369. [CrossRef]
5. Direct Drive Solar Coolers Per Henrik Pedersen, Ivan Katic. 2016. Available online: https://www.dti.dk/_/media/56756_SolarChill_a_solar_pv_refrigerator_without_battery.pdf (accessed on 20 December 2018).
6. Guruprasad, A.; Lingkun, L.; Xiang, H.; Fang, G. Thermal energy storage materials and systems for solar energy applications. *Renew. Sustain. Energy Rev.* **2017**, *68*, 693–706.

7. Sathishkumar, A.; Kathirkaman, M.D.; Ponsankar, S.; Balasuthagar, C. Experimental investigation on solidification behaviour of water base nanocomposite pcm for building cooling applications. *Indian J. Sci. Technol.* **2016**, *9*, 1–7. [CrossRef]
8. Ahammed, N.; Lazarus, G.A.; Titus, J.; Bose, J.R.; Wongwises, S. Measurement of thermal conductivity of graphene–water nanocomposite at below and above ambient temperatures. *Int. Commun. Heat Mass Transf.* **2016**, *70*, 66–74. [CrossRef]
9. Harikrishnan, S.; Deepak, K.; Kalaiselvam, S. Thermal energy storage behavior of composite using hybrid nanomaterials as PCM for solar heating system. *J. Therm. Anal. Calorim.* **2014**, *115*, 1563–1571. [CrossRef]
10. Harish, S.; Daniel, O.; Yasuyuki, T.; Masamichi, K. Thermal conductivity enhancement of lauric acid phase change nanocomposite with graphene nanoplatelets. *Appl. Therm. Eng.* **2015**, *80*, 205–211. [CrossRef]
11. Sakr, M.H.; Abdel-Aziz, R.M.; Ghorab, A.A.E. Experimental and theoretical study on freezing and melting in capsules for thermal storage. *ERJ Soubra Fac. Eng* **2008**, *9*, 48–65.
12. Kaygusuz, K.; Sari, A. Thermal energy storage system using a technical grade paraffin wax as latent heat energy storage material. *Energy Sources* **2005**, *27*, 1535–1546. [CrossRef]
13. Kaygusuz, K. Experimental and theoretical investigation of latent heat storage for water based solar heating systems. *Energy Convers Manag.* **1995**, *36*, 315–323. [CrossRef]
14. Regin, A.F.; Solanki, S.C.; Saini, J.S. Latent heat thermal energy storage using cylindrical capsule: Numerical and experimental investigations. *Renew. Energy* **2006**, *31*, 2025–2041. [CrossRef]
15. Zalba, B.; Sánchez-valverde, B.; Marín, J.M. An experimental study of thermal energy storage with phase change materials by design of experiments. *J. Appl. Stat.* **2005**, *32*, 321–332. [CrossRef]
16. TEAP Energy Products. 2004. Available online: http://www.teappcm.com (accessed on 5 September 2018).
17. Yoon, J.I.; Moon, C.G.; Kim, E.; Son, Y.S.; Kim, J.D.; Kato, T. Experimental study on freezing of water with supercooled region in a horizontal cylinder. *Appl. Eng.* **2001**, *21*, 657–668. [CrossRef]
18. Kalaiselvam, S.; Veerappan, M.; Arul Aaron, A.; Iniyan, S. Experimental and analytical investigation of solidification and melting characteristics of PCMs inside cylindrical encapsulation. *Int. J. Sci.* **2008**, *47*, 858–874. [CrossRef]
19. Novoselov, K.; Geim, A.K.; Morozov, S.V.; Jiang, D.; Zhang, Y.; Dubonos, S.V.; Grigorieva, I.V.; Firsov, A.A. Electric field effect in atomically thin carbon films. *Science* **2004**, *306*, 666–669. [CrossRef] [PubMed]
20. Mehrali, M.; Tahan Latibari, S.; Mehrali, M.; Mahlia, T.M.I.; Metselaar, H.S.C.; Naghavi, M.S. Preparation and characterization of palmitic acid/graphene nanoplatelets composite with remarkable thermal conductivity as a novel shape-stabilized phase change material. *Appl. Eng.* **2013**, *61*, 633–640. [CrossRef]
21. Mehrali, M.; Sadeghinezhad, E.; Latibari, S.T.; Kazi, S.N.; Mehrali, M.; Zubi, M.N.B.M.; Metselaar, H.S.C. Investigation of thermal conductivity and rheological properties of nanofluids containing graphene nanoplatelets. *Nanoscale Res. Lett.* **2014**, *9*, 15. [CrossRef] [PubMed]
22. Liu, Y.D.; Li, X.; Hu, P.F.; Hu, G.H. Study on the supercooling degree and nucleation behavior of water-based graphene oxide nanofluids PCM. *Int. J. Refrig.* **2015**, *50*, 80–86. [CrossRef]
23. Liu, J.; Ye, Z.C.; Zhang, L.; Fang, X.M.; Zhang, Z.G. A combined numerical and experimental study on graphene/ionic liquid nanofluid based direct absorption solar collector. *Sol. Energy Mater. Sol. Cells* **2015**, *136*, 177–186. [CrossRef]
24. Hadadian, M.; Goharshadi, E.K.; Youssefi, A. Electrical conductivity, thermal conductivity, and rheological properties of graphene oxide-based nanofluids. *J. Nanopart. Res.* **2014**, *16*, 2788. [CrossRef]
25. Sudeep, P.M.; Taha-Tijerina, J.; Ajayan, P.M.; Narayananc, T.N.; Anantharaman, M.R. Nanofluids based on fluorinated graphene oxide for efficient thermal management. *RSC Adv.* **2014**, *4*, 24887. [CrossRef]
26. Mehrali, M.; Deghinezhad, E.S.; Latibari, S.T.; Mehrali, M.; Togun, H.; Zubir, M.N.M.; Kazi, S.N.; Metselaar, H.S.C. Preparation, characterization, viscosity, and thermal conductivity of nitrogen doped graphene aqueous nanofluids. *J. Mater. Sci.* **2014**, *49*, 7156–7171. [CrossRef]
27. Liu, J.; Wang, F.X.; Zhang, L.; Fang, X.M.; Zhang, Z.G. Thermodynamic properties and thermal stability of ionic liquid-based nanofluids containing graphene as advanced heat transfer fluids for medium-to-high-temperature applications. *Renew Energy* **2014**, *63*, 519–523. [CrossRef]
28. Selvam, C.; Mohan Lal, D.; Harish, S. Thermal conductivity enhancement of ethylene glycol and water with grapheme nanoplatelets. *Thermochim. Acta* **2016**, *642*, 32–38. [CrossRef]
29. Park, S.; Ruoff, R.S. Chemical methods for the production of graphenes. *Nat. Nanotechnol.* **2009**, *4*, 217–224. [CrossRef] [PubMed]

30. Harr, L.; Gallagher, J.S.; Kell, G.S. *NBS/NRC Steam Tables*; Hemisphere Publishing Corporation: Washington, WA, USA, 1984.

31. Marsh, K.N. *Recommended Reference Materials for the Realization of Physicochemical Propertie*; Blackwell Scientific Publications: Oxford, UK, 1987.

32. Sengers, J.V.; Watson, J.T.R. Improved international formulations for the viscosity and thermal conductivity of water substance. *J. Phys.* **1986**, *15*, 1291. [CrossRef]

33. Rabinataj, A.D.; Hassanzadeh, H.A.; Khaki, M.; Abbasi, M. Unconstrained melting and solidification inside rectangular enclosure. *J. Fundam. Appl. Sci.* **2015**, *7*, 436–451. [CrossRef]

Article

Graphene Nanoplatelets Impact on Concrete in Improving Freeze-Thaw Resistance

Guofang Chen, Mingqian Yang, Longjun Xu, Yingzi Zhang * and Yanze Wang

Department of Civil Engineering, Harbin Institute of Technology, Weihai 264209, China
* Correspondence: zhyz@hit.edu.cn; Tel.: +86-187-6919-6052

Received: 19 July 2019; Accepted: 29 August 2019; Published: 1 September 2019

Abstract: Graphene nanoplatelets (GNP) is a newly nanomaterial with extraordinary properties. This paper investigated the effect of GNP on the addition on freeze–thaw (F–T) resistance of concrete. In this experimental study, water to cement ratio remained unchanged, a control mixture without GNP materials and the addition of GNP was ranging from 0.02% to 0.4% by weight of ordinary Portland cement was prepared. Specimens were carried out by the rapid freeze-thaw test, according to the current Chinese standard. The workability, compressive strength, visual deterioration and mass loss of concrete samples were evaluated. Scanning electron microscopy also applied in order to investigate the micromorphology inside of the concrete. The results showed that GNP concrete has a finer pore structure than ordinary concrete; moreover, the workability of GNP concrete reduced, and the compressive strength of specimens was enhanced within the appropriate range of GNP addition; in addition, GNP concrete performed better than the control concrete in the durability of concrete exposed to F-T actions. Specimens with 0.05% GNP exhibited the highest compressive property after 200 F–T cycles compared with other samples.

Keywords: graphenene nanoplatelets; concrete; freeze-thaw cycles

1. Introduction

Concrete is the most commonly and widely used civil engineering material in the world. However, we all know that concrete has poor features, such as a quasi-brittle nature, prone to crack formation and low tensile strength; these defects seriously reduce the mechanical property and durability of concrete structures. Numerous concrete structures are served in various environments, for example harsh low-temperature circumstance. Freeze–thaw durability of concrete materials is a key aspect affecting the durability and life span of concrete structures in cold areas. Concrete is an inherently defective material; its F–T performance relies predominantly upon the interior structure of the material, such as its porousness, crack, pores types and size, transportation, etc. Many researchers have paid attention to the usage of nanomaterials, for example, nano-kaolinite clay [1], nano-silica [2–4], nano-TiO$_2$ [5], nanoalumina [6] and graphene nanomaterials [7–20], in the last decade. Studies have shown that the additive of these nanomaterials could improve the mechanical property and durability of cementitious materials. Graphene as an emerging Nano material can be added into cement paste to enhance construction sustainability due to their extraordinary performance. Some researchers reported that the reinforcing effects of graphene on cement composites could enhance the physical and mechanical properties of cementitious materials through tests. Pan et al. [10] showed that the compressive and flexural performance of cement paste increased by adding 0.05 wt% graphene oxide (GO), and observed that GO in cement matrix enhanced the surface area of the GO composite; they thought that GO was a potential entrant in cement-based materials. Some studies demonstrated that the addition of GO could promote hydration of cement, accelerate crystallite formation and increase the tightness of cement paste [11]. Du et al. [12] firstly investigated the durability of concrete containing low-cost two-dimensional (2D) Graphene Nanoplatelets (GNP), and evaluated the transport properties of concrete subjected to chloride and water environment, the results showed that GNP could enhance the resistance of concrete to chloride ion and water penetration. Mohammed et al. [15] found that GO addition could modify the microstructure and improve the compressive strength of cement matrix, and reported the GO could improve the freeze-thaw resistance of cement;however, they did not give the optimum dosage of GO. Experimental investigation showed that the addition of graphene oxide nanoplatelets (GOS) and GNP in mortar samples would increase the frost resistance of materials, and the discovered microstructure and modulus biography found that GNP and GOS could greatly refine the microstructure of mortar specimen [16]. Li et al. [17] firstly studied GO aggregates as particle size measurement and found that the aspect ratio of GO aggregates is much higher than that of the original GO nanosheets. Adding graphene and GO sheets to cement paste could improve the mechanical properties of the nanocomposite [20] but can reduce the workability of the nano cement materials [13]. Additionally, adding graphene nano-sheets (GNS) to cement paste could accelerate the hydration reaction and increase the quantity of hydration products, and could also improve chloride penetration resistance of cement paste [19]. Furthermore, adding GNP in cement would reduce the resistivity and make composites obvious pressure sensitivity [18]. Some recent research findings showed that graphene nano-sheets and their derivatives (GND) could modify cement-based materials [21]; we found that most recent attention focuses on the performance of graphene-based materials additions in cementitious materials, which can improve the better performance, such as compressive strength and flexural strength and elastic modulus of paste. Currently, little information is given to the properties of graphene modified concrete; unique experimental work has been done and demonstrated that GNP could improve transport resistance of concrete, but it reported that graphene has no influence on the compressive properties of concrete [12]. This is inconsistent with the effect of graphene on cement paste; therefore, much research about that graphene modified concrete and effects of graphene on durability of concrete should be done.

The objective of the paper is to research the effects of GNP addition on F-T resistance. Experiments were conducted on the durability of concrete including 0%, 0.02%, 0.05%, 0.1%, 0.2%, 0.3% and 0.4% GNP, using the rapid freeze–thaw method. Compressive strength and mass loss were carried out on specimens with and without GNP that were exposed to F–T actions. Moreover, the influence of GNP on the micro-structure of concrete using Scanning electron microscopy (SEM) was characterized.

2. Experimental Program

2.1. Material and Specimen Preparation

In order to prepare cement composites, Ordinary Portland Cement of grade 42.5 was purchased from Weihai Shanshui Cement Co., Ltd. (Weihai, China). Table 1 display the chemical composition of cement. The maximum size of natural sand and coarse aggregate are 5 mm and 20 mm, respectively, and the fineness modulus of sand is 2.7. GNP was supplied from Graphene Suspension (20 wt %), and it had been provided by Weihai hi-tech carbon materials Co., Ltd. in China. The Graphene Suspension can be stably dispersed in ethanol aqueous solution, which consists of graphene layers with thickness of 20 nm and diameter of 10 micrometers. The scanning electron microscope (SEM) images can be shown in Figure 1.

Table 1. Chemical composition of P.O 42.5R cement.

Composition	CaO	SiO_2	Al_2O_3	Fe_2O_3	SO_3	MgO	Loss	Of Ignition
Content (%)	63.79	22.6	4.62	3.26	2.29	1.70	3.14	

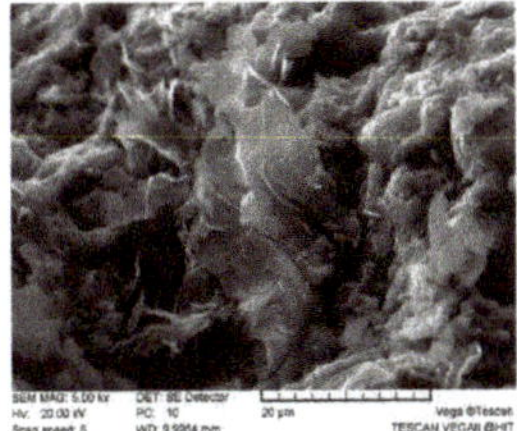

Figure 1. Scanning electron microscope (SEM) images of graphene nanoplatelets (GNP).

In this experiment, to observe the influences of GNP on durability-related behavior, at first, five groups of concrete samples were designed, we defined them as the first batch, which including a control mix and four groups of mixes with 0.1, 0.2, 0.3 and 0.4 wt% GNP, labeled as GC0, GC0.1, GC0.2, GC0.3, and GO0.4, respectively. According to the test results, a dosage of GNP between 0.01% and 0.1% could derive better performance; then, we added two group's specimens and defined as the second batch, 0.02% and 0.05%wt GNP, labeled as GC0.02 and GC0.05. The mass percent of water to cement ratio was kept 0.6, a concrete mix of proportions, by weight of cement, sand and aggregate in this experiment's compound is 1: 2.06: 3.36. Firstly, Graphene Suspension was dissolved in water and manually stirred 1 min; then, cement and graphene-water, which account for about 35% of cement weight, were added to the mixer to stir for 20 s. Afterwards all coarse aggregate are added to stir for 20 s, all sand is added to sir for 20 s, and finally, the remaining graphene-water mixture is added to stir for 30 s. After being blended, the concrete was casted into the cubic mold (100 mm × 100 mm × 100 mm) and then compacted on a vibration table. Each group contains three specimens. Finally, these samples were cured in the laboratory at a temperature 20 ± 1 °C; after 24 ± 2 hours, all samples were demoulded and cured in a concrete standard curing box at a temperature of 20 ± 1 °C and a relative humidity of 95± 1% for 28 days according to the Chinese standard GB/T 50081-2002 [22].

2.2. Methods

According to the Chinese standard GB/T 50082-2009 [23], specimens should be soaked at least 4 h in water before F-T test. After being submerged in the water for 12 h in this experiment, specimens were taken out for weighting and inspection then prepared for the F-T test. The Fully automatic freeze–thaw test box, which meets the procedure requirements of the Chinese standard [23], was used to produce F–T cycles in water. The F–T action consisted of alternatively falling the temperature of the samples from 5 °C to −17 °C and rising it from −17 °C to 5 °C in 4 h. At periodic intervals of different F-T actions (0, 10, 20, 30, 50, 100, 150, 200), the samples were removed from the Cabinet. It is noteworthy that according to the results of the first batch, the F-T cycles of the second batch of specimens are somewhat different from that of the first batch, with 0, 50, 100, 150, and 200 cycles. We took out the removed samples in the natural state of the laboratory for one day, and then performed physical, workability, mass, compressive strength, and scanning electron microscopy (SEM) experiments.

2.2.1. Scanning Electron Microscopy

In order to understand the impacts of GNP on the micro-property of the concrete specimens, a microstructural morphology was conducted on the control concrete and modified samples. A dimension of 4 mm × 4 mm × 4 mm of the specimen was extracted from the control concrete sample. In order to make the concrete conductive, we sprayed 10 nm gold on the surface of the samples before SEM test. Then, the prepared specimens were discovered using the Tescan Vega II SEMsystem. To compare the plain concrete to the GNP concrete, the same operation procedure was conducted on the GNP-modified concrete samples.

2.2.2. Workability

To study the influence of GNP addition on the workability of graphene concrete (GC) mixtures, the slump test conducted in this experiment was in accordance with the Chinese standard GB/T 50080-2016 [21].

2.2.3. Mass Loss

Before and after F-T cycles, we recorded the mass of each sample using an electronic scale with an accuracy of 0.1 g. Then, the mass loss ratio of each specimen Δm_{ni}, an accuracy of 0.01, was defined as follows:

$$\Delta m_{ni} = \left(1 - \frac{m_{ni}}{m_{0i}}\right) \times 100\% \tag{1}$$

where m_{ni} is the mass of the ith sample after nth cycles and m_{0i} is the ith sample' mass before cycles.

The average mass loss ratio of each group of specimens should be the arithmetic average value of the test results of the Δm_{ni} of three specimens as the measured value. When a negative value of a sample test result occurs, the zero should be taken; then, the mass is the arithmetic average value of three specimens. When the difference between the maximum value or minimum value and the median value is more than 1%, this value should be eliminated, and then the arithmetic average value of the remaining two values should be taken as the representative value; when the difference between the maximum value and the minimum value and the median value is all more than 1%, the median value should be taken as the representative value. Therefore, for each group the mass loss ratio Δm_n was defined as follows:

$$\Delta m_n = \frac{\sum\limits_{i=1}^{3} \Delta m_n}{3} \times 100\% \tag{2}$$

2.2.4. Compressive Strength

The compressive test was conducted on an electro-hydraulic servo compressive testing system with a capacity of 1000 kN. According to present the Chinese code [22] for mechanical properties on concrete, the loading rate was 0.5 MPa/s in this experiment. For each group of samples, three specimens were tested. According to the code, the representative strength was taken follow the three rules: (1) the average values of the three samples are taken as the compressive strength values of the each group of specimens; (2) if the difference between the maximum value or the minimum value exceeds 15% of the median value, the maximum value and the minimum value are discarded together, and the median value is taken as the compressive strength value of the group of specimens; (3) if the difference between the two values and the median value is more than 15% of the median value, the test results of this group of specimens are invalid.

3. Results and Discussion

3.1. Scanning Electron Microscopy

A microstructural morphology on the control concrete and graphene concrete specimens was carried out using scanning electron microscopy. Figure 2 shows the SEM micrographs of the specimen before F-T tests. It can be observed that the sample GC0 has larger volume porosity and contains a lot of acicular and rod cement hydration products, such as AFt, AFm, $Ca(OH)_2$, and C-S-H gel. The graphene nano-particles were reported that it could fill the voids and promote the growth of the hydration products, change the shape and size of hydration crystal, but did not change its type through reacting with cement and graphene [9]. As can be seen from Figure 2, compared with the ordinary concrete specimens, the hydration products of the specimens with GNP are more compact and the microstructure is more uniform, the intersecting microcrystals make the mechanical properties of concrete improved obviously, which would change the freeze-thaw resistance of concrete. However, it should be noted that when the content of GNP exceeds 0.3%wt, GNP and other materials of concrete cannot be well mixed, their distribution cannot be uniform, and GNP could be aggregated, which was also found in the process of mixing materials, forming weak areas in the concrete, leading to stress concentration [8]. Therefore, the continuous increase of GNP mass fraction cannot further improve the mechanical strength and F-T resistance of concrete, such as GC0.4. Nevertheless, when the mass fraction of GNP is less, graphene nanoplatelets exist in isolated or small collective forms, with larger particle spacing and less lap joints, which have little effect on the mechanical properties of the concrete. It can be inferred that when the mass fraction of GNP is within a certain range, graphene would plays a role in strengthening and toughening of concrete.

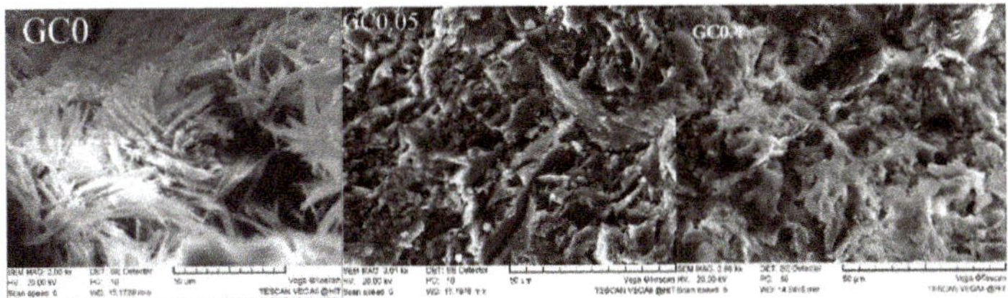

Figure 2. SEM images of the specimens before freeze-thaw tests.

The SEM images of the specimens after 30 F-T actions are shown in Figure 3. Extensive cracking, 0.6–4.7 µm, can be found in the control sample. The GC0.3 and GC0.4 concrete sample exhibited more cracks than the GC0.2 sample, GC0.3 concrete contained 0.4–3.5 µm cracks, but from Table 2 macro-crack cannot be found on these samples surface. Meanwhile, no micro-crack was observed in the GC0.1 concrete sample, this means that the mass fraction of GNP might have a certain range which graphene will play a role in strengthening microstructure of concrete.

After 200 F-T cycles, from the Figure 4 we can observe that the micro-cracks became wide and long (8.6–24.7 µm) in the GC0 samples, tiny micro-cracks began to appear in the GC0.05 (1.3–7.8 µm) and GC0.1 (0.8–9.2 µm) concrete sample. And in the GC0.2, GC0.3 and GC0.4 concrete samples we found that the micro-cracks became wide and more, the crack lengths are 3.4–16.9 µm, 2.7–17.3 µm and 3.3–23.1 µm, respectively, and the internal structure became poriferous, which lessened the durability and mechanical performance of the concrete. This indicates that the perennial F-T cycles could change the microstructure of the concrete and lead to a more porous. The SEM micrographs gives an illustration for that the compressive strength of the GC0.05 sample is maximum in all specimens during the F-T actions.

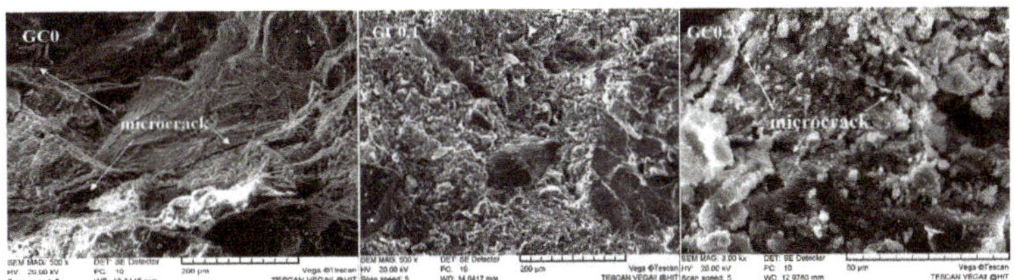

Figure 3. SEM images of the specimens after 30 freeze-thaw tests.

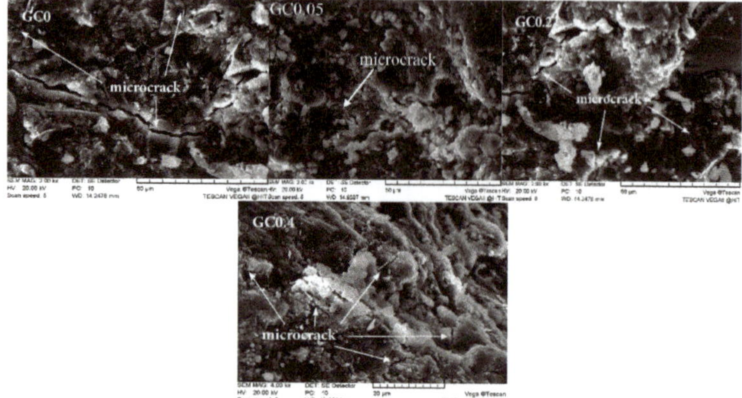

Figure 4. SEM images of the specimens after 200 freeze-thaw tests.

3.2. Workability

The slump results are shown in Figure 5. It can be found that the slump of concrete almost linear reduced with the increase of GNP content. At 0.4 wt % GNP, the mixing process of the cement slurry, stone, and graphene-water mixture became difficult, and the slump was only 57 mm. It is universally acknowledged that GNP has larger surface-to-volume ratio and can absorb more water on its surface; nevertheless, the amount of water added to the fresh slurry remains unchanged during the forming process, and therefore, this could reduce its workability. Meanwhile, with the increase of graphene addition, contact degree magnified between graphene particles; some graphene particles are not uniformly dispersed in cement matrix. This phenomenon can be found in subsequent SEM experiments. Consequently, these would have reduced the concrete workability.

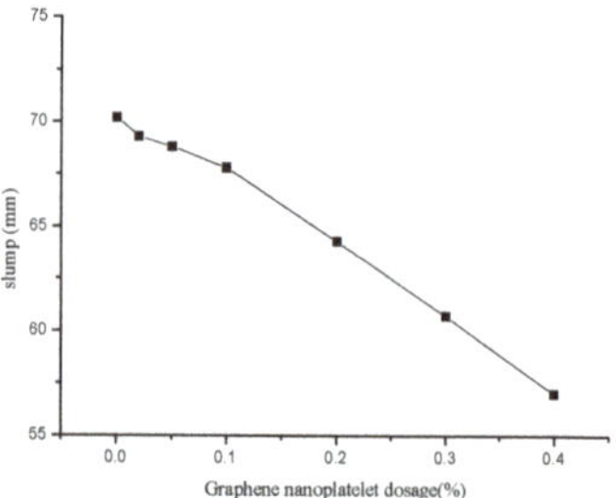

Figure 5. The slump of concrete.

3.3. Visual Assessment

In this experiment, samples were periodically picked out from the device for visual assessment; detailed observations are given in Table 2. With the increase of graphene content, the color of the specimens turns black gradually. It can be noticed that before 20 F-T actions, all specimens with no change in surface. With the increase of freeze-thaw actions, the surface of concrete samples gradually changes. After 30 F-T cycles, tiny cracks and some porous were observed on the surface of control samples, and it was found that the edges of the control specimens began to loosen. However, no obvious change was observed on other GC samples. After 150 F-T cycles, control samples displayed cracking, no damage was observed on GC0.05, GC0.1, and GC0.2 samples, while some porous was found on GC0.02, GC0.3, and GC0.4 samples. After 200 F-T cycles, control samples showed severe cracking and sponge-like surfaces, and GC0.4 became more porous, showing small visible cracking;

however, GC0.05, GC0.1, and GC0.2 still have no obvious damage. From the surface appearance of the sample noticed in the experiment, we can obtain the damage progression. During repeated F-T tests, the water penetrates into the pores of concrete, and when the temperature is lower than a certain value, water will freeze, resulting in expansion stress, resulting in micro-cracks and macro-cracks; therefore, we could observe damages including pores, spalling, and cracks. The addition of GNP causes lower porosity [12] and forms a compact microstructure, as a result of the specimens with GNP might have less water absorption than control specimens, and showed less surface damage compared with the ordinary concrete samples through all the F-T actions.

Table 2. Visual observations of GNP concrete at different F-T cycles.

Cycles	GNP Additives						
	0%	0.02%	0.05%	0.1%	0.2%	0.3%	0.4%
0							
30		—	—				
50							
150							
200							

3.4. Mass Change

Concrete mass could be reduced due to subject to F-T actions. The main reason of mass reduction was involved in the concrete corners due to damage caused by freeze-thaw cycles. Figure 6 shows the mass loss ratio Δm_n after different cycles.

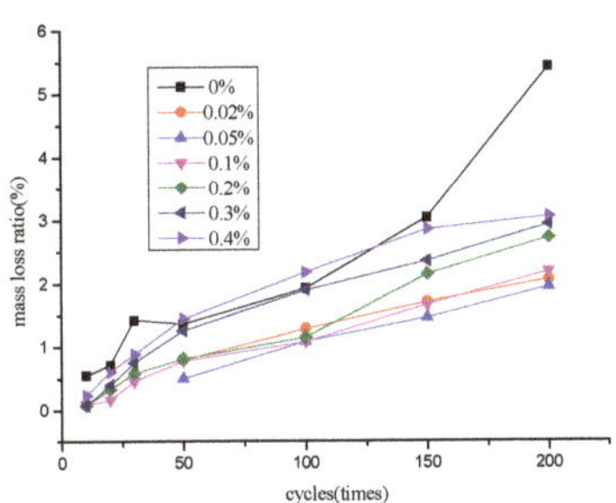

Figure 6. Mass loss ratio of concrete versus F-T cycles.

Although in some F-T cycles (before 50 cycles), the value of the mass loss reduction was small, this phenomenon could be explained as the higher durability of the specimen (such as GC0.05 and GC0.1) against F-T cycles. In freeze-thaw cycle test, mass loss can be used as an index to measure material damage. According to the Chinese standard GB/T 50082-2009 [23], the mass loss of concrete more than 5% means that the material has been destroyed by rapid F-T, and the test can be stopped. From Figure 6, we can see that Δm_n rise gradually for all samples with the cycles continue. After 200 cycles, the maximum value of Δm_n is 5.42% for GC0, followed by 2.06%, 1.96%, 2.19%, 2.72%, 2.92%, and 3.06% for GC0.02, GC0.05, GC0.1, GC0.2, GC0.3, and GC0.4, respectively. This implies that GC0 has been destroyed; therefore, the maximum number of experimental cycles we designed was 200. Moreover, the maximum value of Δm_n of modified concrete implied that GNP has a noticeable ability to keep matrices together; this may be attributed to the smaller size of GNP which could refine the internal pore structure of concrete [12]. It is noteworthy that the mass loss ratio of the concrete samples containing 0.05% GNP is the smallest during the F-T cycles.

3.5. Compressive Strength Evolution

To study the effect of GNP on the compressive strength evolution of the concrete, the compressive strength of the samples of 28day and different F-T cycles was measured. Figure 7a illustrates 28day compressive strength with different GNP additives concrete; it can be found that the concrete compressive strength can be improved by adding GNP up to 0.3%, the enhancement of compressive strength in GC0.02, GC0.05, GC0.1, GC0.2, and GC0.3 relative to that of GC0 was 18.61%, 22.40%, 19.78%, 8.84%, and 1.42%, respectively. When the water cement ratio and coarse aggregate of concrete are the same, the porosity of concrete will greatly affect the compressive strength of samples. There may be two reasons for the increase of strength of the graphene concrete: One is that GNP can accelerate the hydration process of cement [14], which leads to the refinement of pore structure; the other is that graphene nanoplatelets fill the holes in concrete, as evidenced by the SEM results. However, adding 0.4% of GNP had an adverse impact on enhancing the compressive strength of the specimen. The reason for the reduction of 28 days compressive strength in the concrete containing 0.1–0.4% GNP, in comparison with the samples containing 0.05% and 0.02% GNP could be the lack of uniform distribution and nanoplatelets agglomeration and the formation of weak regions in the concrete. The micro structural morphology corroborates the evolution of the strength. However, Du et al. [12] added the GNP in to the concrete, and the GNP content from 0.5% up to 2.5% at 0.5% increment, they reported that GNP has no beneficial or adverse impact on the compressive strength of the concrete, we inferred that they come to this conclusion is that they took a larger increment of graphene content and did not find the appropriate dosage.

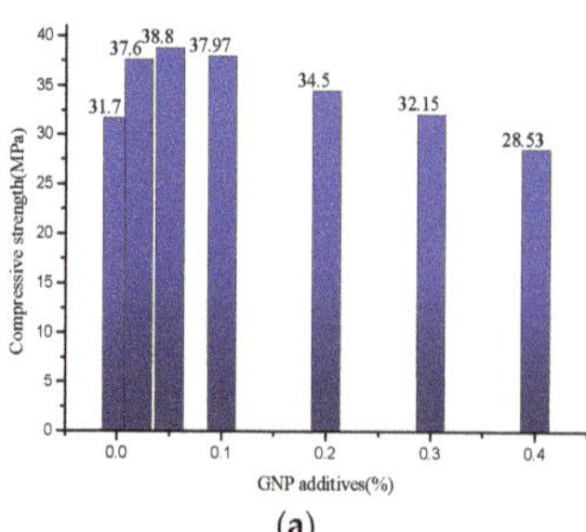
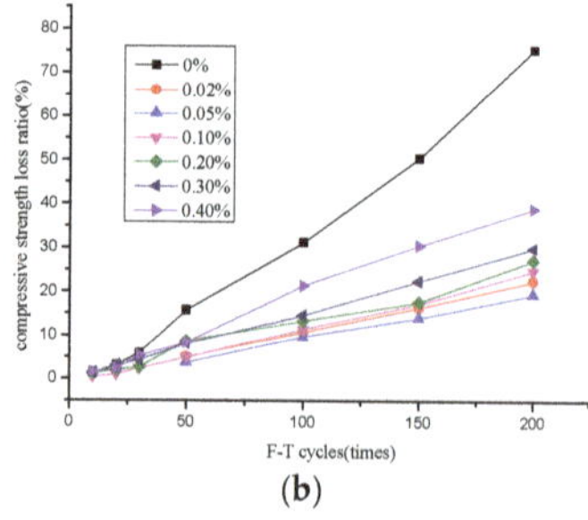

Figure 7. (**a**). Twenty-eight-day compressive strength of different GNP additives. (**b**) D_f versus F-T actions. The compressive strength of concrete with different GNP additives.

To research the changing trend of compressive strength of the sample in tests, compressive strength loss ratio D_f was defined as follows:

$$D_f = \left(1 - \frac{f_n}{f_0}\right) \times 100\% \tag{3}$$

where f_n is the compressive strength of the sample after exposure to the nth cycles, f_0 is the compressive strength of the sample before the experiment. Figure 7b shows the compressive strength loss ratio after F-T actions. The results show that GNP has a remarkable impact on the compressive strength of specimens during the whole test, the compressive strength loss ratio D_f of graphene concrete samples are lower than that of control concrete. After 200 cycles, D_f was 75.39% in the control sample; this means it destroyed according to the China national standard [23], while D_f was 22.60%, 19.59%, 24.85%, 27.25%, 30.02%, and 39.02% in the samples containing 0.02%, 0.05%, 0.1%, 0.2%, 0.3%, and 0.4% GNP. After 200 F-T cycles, the compressive strength of GC0.02 and GC0.05 were 29.1 MPa and 31.2 MPa, respectively. Therefore, adding GNP can significantly improve the durability of concrete under F-T cycles, a probable reason is that there were few or no cracks or fine holes between the crystals in graphene concrete, which greatly reduced the water penetration into the concrete, thus reducing the frozen-heave force; hence, the freeze-thaw resistance of the material was improved. In addition, adding more than 0.05% GNP in this test presented a decreasing trend in improving the compressive strength of graphene specimen in F-T cycles. This means that the mass fraction of GNP is 0.05%wt of the cement; the effect is best in this experiment.

3.6. Elastic Modulus Evolution

The elastic modulus of concrete is usually defined as either a secant or a chord modulus. A secant modulus of concrete is calculated from the starting point to a specified point on the stress–strain curve. The specified point often is selected between 30% and 60% of the specimen's ultimate strength. We take 30% of the ultimate strength of the specimen as the defined point. Table 3 summarizes the elastic modulus of concrete samples. We found that GC0.05 has the strongest ability to resist elastic deformation, and the elastic modulus of GC0.05 is close to that of C60. The dates implied that GNP addition increases modulus of the concrete. It may have contributed to the high stiffness of graphene and the increasing compressive strength. We also found that for all samples, the modulus was obviously changed after 200 cycles. The elastic modulus of graphene concrete showed a lower decrease during the whole test. Comparing with the control sample before F-T actions, the ordinary cement concrete showed a drop up to 25.4% in the elastic modulus at 200 cycles. Nevertheless, the sample with the additives of 0.05%, 0.2%, and 0.3% GNP showed 17.57%, 19.7%, and 16.4% decrease in elastic modulus, respectively. After 200 F-T cycles, the elastic modulus of GC0.05 is still close to 30.0GPa. This indicated that the addition of GNP can enhance the elastic deformation resistance of concrete in F-T actions.

Table 3. Elastic modulus of concrete samples after F-T cycles (unit: GPa).

F-T Cycles	GC0	GC0.02	GC0.05	GC0.1	GC0.2	GC0.3	GC0.4
0	30.895	35.593	36.060	35.485	33.740	31.900	30.355
10	29.445	-	-	34.010	33.490	31.445	29.830
20	29.375	-	-	32.415	33.215	30.655	29.395
30	28.265	-	-	32.790	31.840	30.835	29.190
50	28.295	32.019	33.420	31.755	30.410	30.430	27.550
100	26.700	31.440	32.670	30.380	29.865	28.575	25.560
150	25.175	29.375	30.393	28.825	28.210	27.835	25.060
200	23.045	28.460	29.725	28.140	27.080	26.665	23.315

3.7. Peak Strain Change

In this study, we defined peak strain as the strain when the compressive stress reaches the maximum value in the experiment. Table 4 summarizes the peak strain of concrete samples after different F-T actions. Before cycles the peak strain of GC0 is about 0.00188, and the peak strain of graphene concrete is slightly larger than that of the control concrete, this means that when the specimen is destroyed the graphene concrete has large deformation than the ordinary concrete; a probable reason is that GNP improves the deformability of concrete. From Table 4, we can observe that the peak strain decrease with cycle times and the changing trends for different wt% GNP concrete with different F-T cycles is similar.

Table 4. Peak strain of concrete after F-T cycles.

F-T Cycles	GC0	GC0.02	GC0.05	GC0.1	GC0.2	GC0.3	GC0.4
0	0.00188	0.00224	0.00232	0.00229	0.00220	0.00209	0.00189
10	0.00194	-	-	0.00236	0.00214	0.00203	0.00185
20	0.00178	-	-	0.00217	0.00214	0.00199	0.00180
30	0.00175	-	-	0.00195	0.00207	0.00195	0.00184
50	0.00173	0.00187	0.00192	0.00185	0.00184	0.00174	0.00167
100	0.01051	0.00163	0.00171	0.00161	0.00158	0.00154	0.00158
150	0.00152	0.00156	0.00166	0.00157	0.00161	0.00151	0.00154
200	0.00139	0.00152	0.00164	0.00151	0.00147	0.00144	0.00133

4. Conclusions

This paper researched the effect of GNP on the freeze-thaw performance of concrete. A series of experimental investigations were carried out. Concrete samples consist of 0%, 0.02%, 0.05%, 0.1%, 0.2%, 0.3%, and 0.4% graphene nanoplatelets were prepared for the rapid F-T tests. The workability, compressive strength, and durability graphene of concrete subjected to F-T cycles of the specimens were studied, and the main conclusions were as follows.

(1) The SEM micrographs indicate that the effect of GNP enhanced the freeze-thaw resistance of concrete. It can be concluded that graphene nanoplatelets with the appropriate additional acts effectively in enhancing the durability of concrete under freezing-thawing cycles.

(2) Adding graphene nanoparticles could reduce the workability of graphene concrete, and the slump of the specimen decreases with the increasing nanoparticles addition.

(3) Concrete with GNP displayed less mass loss during F-T actions, particularly the concrete with the 0.05% GNP addition. In addition, GNP concrete exhibited fewer damages on the sample's surfaces than the control concrete.

(4) Adding graphene nanoparticles up to 0.3% could improve the compressive strength of concrete; before F-T test, the maximum increase of compressive strength in GC0.05 relative to that of GC0 was 22.40%. However, above 0.4%, the incorporation of GNP had a negative influence on enhancing the compressive strength of concrete. Furthermore, concrete with the appropriate additional GNP showed less compressive strength loss after 200 F-T cycles compare to the control specimens.

Author Contributions: G.C. wrote the manuscript, Y.Z. designed the theoretical framework, M.Y. and Y.W. performed the experiments, L.X. corrected the language.

Funding: This research was funded by the National Natural Science Foundation of PR. China (Grant No.51308166), Shandong Provincial Natural Science Foundation (Grant No. ZR2019MEE090) and Weihai Science and technology Development Plan Project (Grant No.2015DXGJMS011).

Acknowledgments: We thank the Analysis and Testing Center of Harbin University of Technology (Weihai) for assistance with the SEM analysis.

Conflicts of Interest: The authors declare no conflict of interest.

References

1. Fan, Y.; Zhang, S.; Wang, Q.; Shah, S.P. Effects of nano-kaolinite clay on the freeze–thaw resistance of concrete. *Cem. Concr. Compos.* **2015**, *62*, 1–12. [CrossRef]
2. Du, H.; Du, S.; Liu, X. Durability performances of concrete with nano-silica. *Constr. Build. Mater.* **2014**, *73*, 705–712. [CrossRef]
3. Mukharjee, B.B.; Barai, S.V. Statistical techniques to analyze properties of nano-engineered concrete using Recycled Coarse Aggregates. *J. Clean. Prod.* **2014**, *83*, 273–285. [CrossRef]
4. Stynoski, P.; Mondal, P.; Marsh, C. Effects of silica additives on fracture properties of carbon nanotube and carbon fiber reinforced Portland cement mortar. *Cem. Concr. Compos.* **2015**, *55*, 232–240. [CrossRef]
5. Li, H.; Xiao, H.; Guan, X.; Wang, Z.; Yu, L. Chloride diffusion in concrete containing nano-TiO$_2$ under coupled effect of scouring. *Compos. Part. B Eng.* **2014**, *56*, 698–704. [CrossRef]
6. Shahrajabian, F.; Behfarnia, K. The effects of nano particles on freeze and thaw resistance of alkali-activated slag concrete. *Constr. Build. Mater.* **2018**, *176*, 172–178. [CrossRef]
7. Lv, S.; Ma, Y.; Qiu, C.; Sun, T.; Liu, J.; Zhou, Q. Effect of graphene oxide nanosheets of microstructure and mechanical properties of cement composites. *Constr. Build. Mater.* **2013**, *49*, 121–127. [CrossRef]
8. Du, H.; Pang, S.D. Enhancement of barrier properties of cement mortar with graphene nanoplatelet. *Cem. Concr. Res.* **2015**, *76*, 10–19. [CrossRef]
9. Mingli, C.; Huixia, Z.; Cong, Z. Effect of graphene on mechanical properties and microstructure of cement paste. *J. Harbin. Inst. Technol.* **2015**, *47*, 26–30.
10. Pan, Z.; He, L.; Qiu, L.; Korayem, A.H.; Li, G.; Zhu, J.W.; Collins, F.; Li, D.; Duan, W.H.; Wang, M.C. Mechanical properties and microstructure of a graphene oxide–cement composite. *Cem. Concr. Compos.* **2015**, *58*, 140–147. [CrossRef]
11. Wang, Q.; Wang, J.; Lu, C.-X.; Liu, B.-W.; Zhang, K.; Li, C.-Z. Influence of graphene oxide additions on the microstructure and mechanical strength of cement. *New Carbon Mater.* **2015**, *30*, 349–356. [CrossRef]
12. Du, H.; Gao, H.J.; Pang, S.D. Improvement in concrete resistance against water and chloride ingress by adding graphene nanoplatelet. *Cem. Concr. Res.* **2016**, *83*, 114–123. [CrossRef]
13. Gong, K.; Pan, Z.; Korayem, A.H.; Qiu, L.; Collins, F.; Wang, C.M.; Duan, W.H. Reinforcing Effects of Graphene Oxide on Portland Cement Paste. *J. Mater. Civ. Eng.* **2016**, *27*, A4014010. [CrossRef]
14. Ming-li, C.; Hui-xia, Z.; Cong, Z. Effect of graphene on mechanical properties of cement mortars. *J. Cent. South Univ.* **2016**, *23*, 919–925.
15. Mohammed, A.; Sanjayan, J.G.; Duan, W.H.; Nazari, A. Graphene Oxide Impact on Hardened Cement Expressed in Enhanced Freeze–Thaw Resistance. *J. Mater. Civ. Eng.* **2016**, *28*, 1–6. [CrossRef]
16. Tong, T.; Fan, Z.; Liu, Q.; Wang, S.; Tan, S.; Yu, Q. Investigation of the effects of graphene and graphene oxide nanoplatelets on the micro- and macro-properties of cementitious materials. *Constr. Build. Mater.* **2016**, *106*, 102–114. [CrossRef]
17. Li, X.; Lu, Z.; Chuah, S.; Li, W.; Liu, Y.; Duan, W.H.; Li, Z. Effects of graphene oxide aggregates on hydration degree, sorptivity, and tensile splitting strength of cement paste. *Compos. Part. A Appl. Sci. Manuf.* **2017**, *100*, 1–8. [CrossRef]
18. Xu, J.; Zhang, D. Pressure-sensitive properties of emulsion modified graphene nanoplatelets/cement composites. *Cem. Concr. Compos.* **2017**, *84*, 74–82. [CrossRef]
19. Wang, B.; Zhao, R. Effect of graphene nano-sheets on the chloride penetration and microstructure of the cement based composite. *Constr. Build. Mater.* **2018**, *161*, 715–722. [CrossRef]
20. Liu, J.; Li, Q.; Xu, S. Reinforcing Mechanism of Graphene and Graphene Oxide Sheets on Cement-Based Materials. *J. Mater. Civ. Eng.* **2019**, *34*, 1–9. [CrossRef]
21. Yang, H.; Cui, H.; Tang, W.; Li, Z.; Han, N.; Xing, F. A critical review on research progress of graphene/cement based composites. *Compos. Part. A Appl. Sci. Manuf.* **2017**, *102*, 273–296. [CrossRef]

22. China National Standard. *Standard for Test Methods of Mechanical Properties on Ordinary Concrete (GB/T50081-2002)*; China Architecture and Building Press: Beijing, China, 2002.
23. China National Standard. *Standard for Test Methods of Long-Term Performance and Durability of Ordinary Concrete (GB/T 50082-2009)*; China Architecture and Building Press: Beijing, China, 2009.

Article

Facile Preparation of Graphene Oxide-MIL-101(Fe) Composite for the Efficient Capture of Uranium

Bing Han *, Enyao Zhang and Gong Cheng

MOE Key Laboratory of Resources and Environmental Systems Optimization, College of Environmental Science and Engineering, North China Electric Power University, Beijing 102206, China; zhangenyao0516@163.com (E.Z.); cg1182229006@ncepu.edu.cn (G.C.)
* Correspondence: hanbing01@ncepu.edu.cn; Tel.: +86-10-6177-1470

Received: 11 October 2018; Accepted: 13 November 2018; Published: 16 November 2018

Abstract: Graphene oxide (GO)-MIL-101(Fe) (Fe-based metal-organic frameworks (MOFs) with Fe(III) as the metal anode and 2-aminobenzene-1,4-dicarboxylic acid as a ligand) sandwich composites are designed and fabricated through a facile in situ growth method. By modulating the addition amount of GO nanosheets, composites containing MIL-101(Fe) octahedrons with a tunable dimension and density are achieved. The optimized ratio between individual components is determined through adsorption experiments. Adsorption isotherms reveal an enhanced adsorption efficiency and improved adsorption capacity of GO15-MIL-101(Fe) (GO dosage is 15 mg) in comparison with raw MIL-101(Fe) nanocrystals. Experimental evidence indicates that the removal of U(VI) by the composite is based on inner-sphere surface complexation and electrostatic interaction. The improved adsorption performance originates from the optimized synergistic effects of GO and MIL-101(Fe) octahedrons. In summary, this work offers a facile synthetic method to achieve cost-effective composites towards the U(VI) capture. It also lays the foundation for the design of novel adsorbents with the full play of component's functionality.

Keywords: graphene oxide; MIL-101(Fe); composite; adsorption; uranium

1. Introduction

The severe energy and environmental crisis calls for the development of green energy sources. Nuclear power is an effective solution for this problem as a sustainable energy source with a high efficiency. However, the radioactive contamination from the production process of nuclear energy can cause potential ecological threats and biological toxicity [1–3]. In particular, uranium-contamination is included in the major nuclide pollutants that are urgently required to be eliminated from the polluted waters [4]. The widely used techniques for the removal of uranyl ions consist of solvent extraction [5], ion-exchange [6], and adsorption. In particular, adsorption takes advantages of the rapid and highly efficient treatment of uranium by developing adsorbent materials with programmable functionalization [7]. Until now, much attention has been paid to the development of nanostructured adsorbents for higher adsorption efficiency [8,9]. One of the most promising directions is the fabrication of composite materials which combine the functionality of individual components into one [10,11]. The key point lies in the component selection and sophisticated structure design for the optimization of synergistic effect inside the composite materials.

Among the various adsorbents, functional carbon nanomaterials including one dimensional carbon nanofiber/nanotube, two-dimensional graphene oxide (GO) [12,13], and three dimensional functionalized carbon nanostructures take the advantages of a large surface area and easy functionalized capability, which endow them the great potential as novel adsorbents for uranium-based nuclide species [14–18]. Zänker et al. reported that the adsorption of uranyl ions on pristine carbon nanotube (CNT) reached 5.0 mmol g^{-1} at pH = 5.0 under 298 K [19]. Wang et al. demonstrated the good adsorption performance of GO towards U(VI) in aqueous solution [20]. The tedious and high-cost synthetic procedures of CNT and GO promote the studies on the fabrication of CNT/GO-based composites which maintain a good performance at a lower dosage of CNT and GO by the introduction of other components and optimizing the synergistic effects [21,22]. Compared with CNT, fabrication of GO-based composite materials benefits the adsorption properties in several ways. On the one hand, the expansive surface of GO gives rise to the easy attachment of other nanomaterials, making them excellent substrates. On the other hand, the formation of composites through surface growth method helps overcome the aggregation problem of GO in solutions, which improves the exposure of functional groups. In the past few years, various GO-based composite adsorbents with diverse functionalities have been designed and fabricated, including a magnetic GO composite [23], GO-polymer composite [24,25], GO-layered double hydroxides (LDHs) composite [26], etc.

Metal-organic frameworks (MOFs) are a series of crystalline hybrids based on the bridging network between metal ions and organic ligand molecules [27]. Owing to their high surface area, tunable porous structure, and versatile chemical functionality, MOFs have become high-profile candidates in application fields of catalysis [28], adsorption [29–32], and membrane formation [33]. Very recently, MOFs have found new opportunities in the removal of uranyl ions [34,35]. MIL-101, as a series of MOFs with different metal centers (Al, Cr, and Fe), has received much attention in the recovery of uranium polluted waters due to its chemical stability in acid solutions [36–38]. Compared with pristine materials, functionalized MOFs have shown optimized adsorption performance with enriched binding sites [39]. The post-synthetic strategy is the commonly used method to introduce functional groups on the MIL-101 host. However, the multistep post-grafting process is time-consuming and usually involves polluting acid such as HNO_3 [40]. One step or the in situ fabrication of MOF-based composite is a more facile and environmentally benign synthetic method to achieve novel adsorbents with a satisfying performance. Several previous studies have provided excellent examples for GO-MOF composites as adsorbents [41,42].

Herein, we presented the fabrication of the GO-MIL-101(Fe) composite through in situ growth method. Different ratio of GO and MIL-101(Fe) in the composite was tuned through changing the addition amount of GO nanosheets. The size and distribution density of MOF polyhedrons varied due to the interference of the functional groups on GO nanosheets towards the nucleation and growth of MIL-101 seed. Morphology imaging based on scanning electron microscopy (SEM) and a composition analysis from powder X-ray diffraction (XRD), Attenuated total internal reflection Fourier (ATR-FTIR) and X-ray photoelectron spectroscopy (XPS) measurements eVidenced the successful synthesis of the composite material. Batch adsorption experiments further confirmed the much better adsorption performance of the GO15-MIL-101(Fe) composite than pristine MIL-101(Fe) with larger U(VI) adsorption capacity and superior reusing ability. Mechanism analysis further illustrated the importance of optimizing synergistic effects between individual components in improving the adsorption performance. This work would benefit the design and development of composite adsorbents with a versatile functionalization for the potential application in the effective recovery of U(VI) pollution.

2. Materials and Methods

2.1. Chemicals

Graphite (CP grade, ≤300 mesh) was purchased from Sinopharm Chemical Regent, Beijing, China. Iron(III) chloride hexahydrate ($FeCl_3 \cdot 6H_2O$, 97 wt%), 2-aminobenzene-1,4-dicarboxylic acid (NH_2-H_2BDC, 98 wt%), and *N,N*-dimethylformamide (DMF, 99.8 wt%) were provided by Alfa Aesar, Shanghai, China. Polyvinylpyrrolidone (PVP, Mw = 30,000) was provided by Xilong Scientific Co., Ltd., Shanghai, China. Uranyl nitrate was obtained from Sigma Aldrich, Beijing, China. The graphene oxide (GO) was synthesized based on the Hummers method [43].

2.2. Characterization

The dimension and morphology of all the samples were obtained from scanning electron microscopy (SEM) (Hitachi S4800, Tokyo, Japan). X-ray diffraction (XRD) D/max-TTRIII (Rigaku, Tokyo, Japan) was used to provide the crystalline information of the samples. Attenuated total internal reflection Fourier (ATR-FTIR) spectra were obtained on a Bruker TENSOR-27 (Bruker Optics, Bremen, Germany). The surface composition and binding state information were achieved from X-ray photoelectron spectroscopy (XPS) (Thermo Escalab 250Xi, Waltham, MA, USA). Pore information and specific surface area were analyzed by a Micromeritics ASAP 2010 (Norcross, GA, USA). ζ potential data were collected from a Zetasizer Nano ZS (Malvern Instrument, Worcestershire, UK). For measuring the zeta potential, the pH values of the achieved samples (c_{NaCl} = 10 mM) were adjusted to values between 1 and 11 by adding HCl or NaOH, respectively.

2.3. Fabrication of the GO-MIL-101(Fe) Composite with Different GO Contents

GO-MIL-101(Fe) was synthesized with an in situ growth method. In a typical procedure, 10 mg, 15 mg, 18 mg, 20 mg of GO and 54.1 mg of $FeCl_3 \cdot 6H_2O$ (0.2 mmol) were dispersed in 6 mL DMF and formed solution A, 18.2 mg NH_2-H_2BDC (0.1 mmol) and 4 mg PVP were dissolved in 6 mL DMF to form solution B. Subsequently, solutions A and B were placed in a microwave vessel and sealed. After reaction at 160 °C for 10 min, the solid products were collected by centrifugation at 5000 rpm for 10 min. The samples were washed twice with DMF followed by washing with ethanol two times to remove residual DMF. Finally, the product was dried in vacuum overnight for further use. The achieved composite was denoted as GO10(15, 18, 20)-MIL-101(Fe). MIL-101(Fe) materials were fabricated according to the above method without the addition of GO. The mass ratio of GO and MIL-101(Fe) was calculated to be 1:7.42 by measuring the mass of both GO15-MIL(101) and the original added GO.

2.4. Batch Sorption Experiments

The adsorption properties of composite samples were studied by batch experiments. Stock dispersions of adsorbents including MIL-101(Fe) and the GO10(15, 18, 20)-MIL-101(Fe) composites were prepared with a concentration of 1.2 g L^{-1}. U (VI) stock solution (0.2 g L^{-1}) was also prepared. Adsorption suspensions to perform adsorption experiment with different conditions were prepared by diluting the stock adsorbent suspension and U(VI) solution with distilled water. The adsorbent concentration in the adsorption suspension was 0.20 g L^{-1} except for the adsorption experiment with different adsorbent doses. The initial U concentration of 10 mg L^{-1} is selected to study the adsorption behavior of the GO-Fe MOF composite towards U(VI), which simulated the example of industrial effluents with a relatively high U concentration. The effect of pH on the adsorption efficiency was investigated by tuning the solution pH with a different concentration of HNO_3 or NaOH (0.01 mM~0.1 M), which guaranteed that the added amount of HNO_3 or NaOH was almost negligible in comparison with the volume of the adsorption solution. The pH modulation process was monitored by a pH meter. The adsorption efficiency of GO15-MIL-101(Fe) under different initial U concentrations (0.035 mg L^{-1}, 1.3 mg L^{-1}, 3 mg L^{-1}) was also tested (pH = 5.5, T = 298 K). These concentrations took reference from a previous study [6]. The adsorption isotherms were achieved by performing adsorption experiment under different initial U(VI) concentrations (5 mg L^{-1}~70 mg L^{-1} with 5 mg L^{-1} intervals) under pH = 5.5 at different temperatures (298 K, 313 K, 328 K). The pH of the adsorption suspensions was tuned to 5.5 at 298 K and then the adsorption solutions were placed in a shaker under different temperatures. Adsorption kinetics were performed at ambient temperature (pH = 5.5). The influence of coexisting ions (cations from $NaNO_3$, KNO_3, $Mg(NO_3)_2$, $Ca(NO_3)_2$ and anions from $NaNO_3$, NaCl, $Na(SO_4)_2$, and $Na(CO_3)_2$, the concentration was 0.01 M) towards the adsorption performance was further investigated. Particularly, the removal efficiency of U(VI) on GO15-MIL-101(Fe) and MIL-101(Fe) was compared in simulated surface water and groundwater.

The pH, composition, and related ion concentrations were determined based on previous studies (Table S1) [44]. The suspension was agitated for over 24 h to achieve equilibrium. The concentration of uranyl ions under equilibrium was confirmed by the Arsenazo III based colorimetric method. U (VI) can form a binary complex with Arsenazo III. The uranyl ion content was determined by measuring the absorbance of the complex at 656 nm. Samples with U(VI) concentration in ppb level were tested via inductively coupled plasma optical emission spectrometry (ICP-OES) analysis. The adsorption percentage (Adsorption, %) and adsorption capacity (q_e, mg g^{-1}) can be obtained based on the following equations:

$$\text{Adsorption} = \frac{(C_0 - C_e)}{C_0} \times 100\% \tag{1}$$

$$q_e = \frac{(C_0 - C_e) \times V}{m} \tag{2}$$

where, C_0 (mg L^{-1}) and C_e (mg L^{-1}) correspond to the U(VI) concentration of the adsorption suspension at the beginning and equilibrium, respectively. V (mL) and m (g) are the solution volume and adsorbent dosage, respectively. Triplicate measurements showed that the relative errors were within 5%.

2.5. Desorption and Reusing Studies

The reusing ability of the GO15-MIL-101(Fe) was studied through monitoring the adsorption efficiency of the 8-cycles regenerated samples. U(VI) was desorbed from the sample through the washing process of 0.1 M HNO$_3$. The as regenerated sample was centrifuged and washed with deionized water for the removal of acid. Then the dried powder was used in the next adsorption process. The U containing HNO$_3$ wastewater was collected by a professional liquid waste treatment plant or stored to be reused as raw pollutant water in further U removal research.

3. Results and Discussion

3.1. Characterization of GO-MIL-101(Fe) Composite

As shown in Figure 1a, pure GO presented a two-dimensional morphology with lateral size up to tens of micrometers. GO-MIL-101(Fe) composites were fabricated through a facile one-pot in situ growth method. The component ratio was adjusted by tuning the addition amount of GO in the synthetic precursor solution. Figure 1b–d confirmed the successful growth of nanosized MIL-101(Fe) octahedrons on GO nanosheets. The negligible dispersed MIL-101(Fe) indicated that most MIL-101(Fe) nanoparticles formed on the GO matrix. Moreover, MIL-101 octahedrons were well distributed on both sides of the GO nanoplatelets, which guaranteed the synergistic interaction between individual compositions. Notably, the dimension of MIL-101(Fe) octahedrons in the composite decreased with the amount of GO increasing (460 nm for GO10-MIL-101(Fe), 350 nm for GO15-MIL-101(Fe), and 150 nm for GO20-MIL-101(Fe)). It was reported that low concentrations of a monocarboxylic acid with different pKa values (1.60~10.15) could modulate the crystal growth of MIL-101(Cr), with higher pKa leading to smaller nanoparticles [45]. GO synthesized from Hummers method possessed many oxygenated groups like carboxylic (pKa = 4.3 and 6.6) and phenolic groups (pKa = 9.8) [46]. Therefore, both carboxylic groups and phenolic groups gave rise to the controlling of the size of MIL-101 in GO15-MIL-101(Fe) composite. Considering that most MIL-101 octahedrons nucleated on the surface of GO, phenolic groups with especially high pKa played a dominant role. We postulated that GO interfered with the nucleation of MIL-101(Fe) which led to smaller sized nanoparticles with the addition amount of GO increasing.

As shown in Figure S1, all the XRD patterns of the composite materials showed peaks at 2θ = 5.88° and 9.08°, which can be assigned to typical (531), (911) diffractions of MIL-101(Fe), confirming the successful synthesis of the composite sample [47,48]. However, the decreased peak intensity in GO20-MIL-101(Fe) indicated the reduction of MOF crystallinity, which was also reflected from the loss of the octahedron morphology. ATR-FTIR spectroscopy was used to identify the functional groups of the GO-MIL-101(Fe) composite. The broad band in the range of 3000 cm^{-1}~3500 cm^{-1} originated from the hydroxyl groups of absorbed water molecules. The IR signals at 3475 cm^{-1} and 3345 cm^{-1} in Figure 2a corresponded to the asymmetric and symmetric N–H stretching vibrations in amino groups of MIL-101(Fe) [49]. Because of the low concentration of MIL-101(Fe) in the composite, these peaks became negligible in Figure 2b,c. For the sample MIL-101, FT-IR signals at 1575 cm^{-1} and 1429 cm^{-1} were asymmetric and symmetric stretching of carboxylic groups in the organic ligand NH$_2$BDC, respectively [50,51]. The peak at ~1379 cm^{-1} obviously indicated the effect of the amino groups on the symmetric stretching of carboxylic groups. C–N-stretching vibrations were clearly recognized at 1257 cm^{-1} in IR spectrum of MIL-101(Fe) [40]. The spectrum peaks at 1501 cm^{-1} and 767 cm^{-1} can be attributed to the stretching vibration (C=C) and deformation vibration (C–H) of benzene, respectively [52]. As shown in Figure 2b, the asymmetric stretching of O=C–O, C–N stretching, and C–H deformation vibration were suppressed, which was more obvious with the GO increasing. This result implied the modulation of GO towards the growth of MIL-101(Fe). The comparison of FTIR spectra between GO15-MIL-101(Fe) and that loaded with uranyl ions can be used to deduce the uranyl ions-composite interaction. As shown in Figure 2c, the peak at ~920 cm^{-1} can be assigned to the stretching vibration of the linear structure of the UO$_2$$^{2+}$ group in the U(VI)-loaded materials [53].

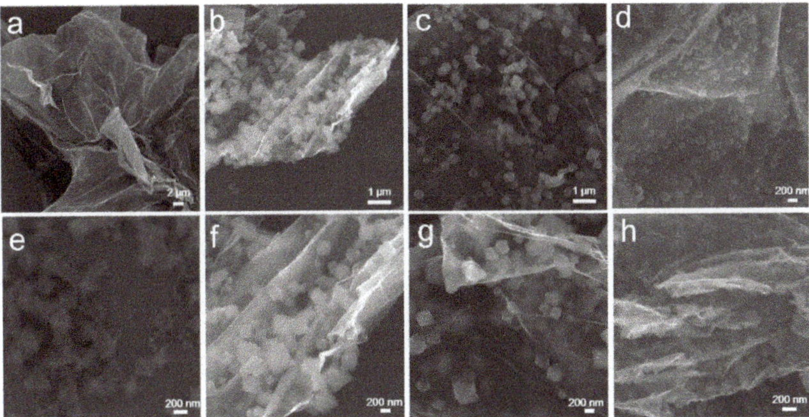

Figure 1. The scanning electron microscopy (SEM) images of (**a**) GO, (**b**,**f**) GO10-MIL-101(Fe) composite, (**c**,**g**) GO15-MIL-101(Fe) composite, (**d**,**h**) GO20-MIL-101(Fe) composite and (**e**) MIL-101(Fe).

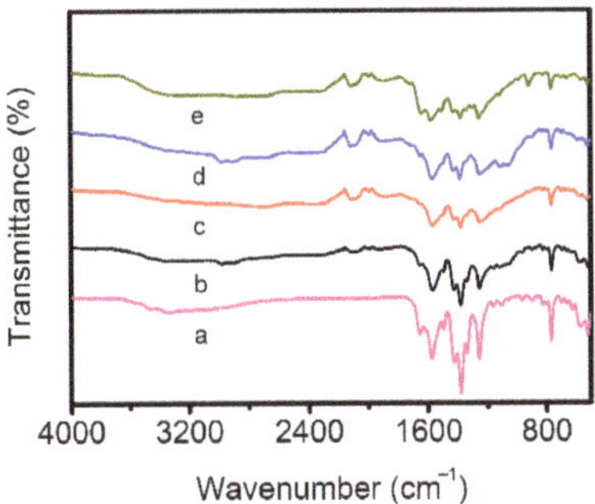

Figure 2. The Attenuated total internal reflection Fourier (ATR-FTIR) spectra of (a) MIL-101(Fe), (b) GO10-MIL-101(Fe), (c) GO15-MIL-101(Fe), (d) GO20-MIL-101(Fe) and (e) GO-15-MIL-101(Fe) loaded with U(VI).

The XPS measurement provided the information of elemental composition and surface binding condition of GO-MIL-101(Fe) composite. XPS survey spectra showed the typical peaks of Fe2p, C1s, O1s, and N1s in GO-MIL-101(Fe) composites and additional U4f peaks upon the uptake of uranyl ions (Figure 3a). The O auger signal (976 eV) was observed [54]. The Fe auger signals (893 eV, 841 eV, and 786 eV) and U4d$_{5/2}$ signal (740 eV) [55,56] were also recognized. XPS signals at 311 eV and 554.9 eV can be assigned to the C1s and O1s satellite peaks [57–59]. The negligible observation of the characteristic signal of DMF molecules (C–N symmetric stretching at 866 cm^{-1} and the O=C–N stretching mode at 659 cm^{-1}) indicated the removal of residual DMF molecules in the system. The C1s spectra of the composite materials can be deconvoluted into three peaks including C–C, C–O, and C=O with the binding energy increasing. O1s spectra were composed of oxygen anion (OH^{-1}) at 530.9 eV, C=O (carboxylic group) at 531.5 eV and C–O in aromatic rings, phenols, and ethers at 533.4 eV [60]. The analysis of the XPS data for GO-MIL-101(Fe) samples before and after the uranyl ions adsorption demonstrated the interaction between the adsorbents and U(VI). As shown in Figure 3b, GO-MIL-101(Fe)-U(VI) exhibited typical U4f XPS peak which consisted of two doublet-peaks at 382 eV (U4f$_{7/2}$) and 392 eV (U4f$_{5/2}$) [6]. The peak at 399.97 eV and 403.13 eV corresponded to the N1s XPS signal with the former corresponding to the amine groups on MIL-101(Fe) and the latter peak indicated the presence of N doping in the form of pyridinic N-oxide on GO. Such N doping may originate from the DMF treatment under the solvothermal condition [61]. The decreased intensity of the N1s peak in GO-MIL-101(Fe)-U in comparison with the raw sample indicated that the amino group of NH$_2$BDC ligand contributed partial adsorption performance. The decreased intensity of C–O and C=O peaks of C1s as well as the decreased area of O1s peaks indicated the interaction between the uranyl ions and different sort of functionalized oxygen groups on the composite. The XPS analysis eVidenced the synergistic effect of GO and MIL-101(Fe) in the removal of U(VI), which agrees well with the FT-IR results.

Brunauer–Emmett–Teller (BET) analysis further provided the surface area and pore structure information. MIL-101(Fe) showed a mixture of type I and IV adsorption-desorption isotherms with the sharp N$_2$ uptakes at low relative pressure (P/P$_0$ < 0.1), indicating its micro/mesoporous structure (Figure S2a) [62]. The pore size distribution result (Figure S2b) indicated the existence of 2.15 nm and 2.52 nm mesocages and 1.26 nm microcages in MIL-101(Fe). In contrast, GO-15-MIL-101(Fe) also exhibited characteristics of type II isotherms, which revealed that the introduction of macropores in the composite (Figure S2c). The H$_3$-type hysteresis loops implied that the sandwich structures formed in the 2-D GO nanosheets-MIL-101 octahedrons composite gave rise to slit mesopores [47,63]. Figure S2d further showed enlarged mesocages of ~2.75 nm. Both isotherm observation and statistic data (Table S2) showed that GO15-MIL-101(Fe) had a lower specific surface area and larger total pore volume than that of pure MIL-101(Fe), which resulted from the formation of meso-macropores in the composite structure. The pore texture analysis illustrated that the GO15-MIL-101(Fe) composite synthesized from the in situ growth method may modulate the pore structure of both individual components, which may optimize the adsorption performance.

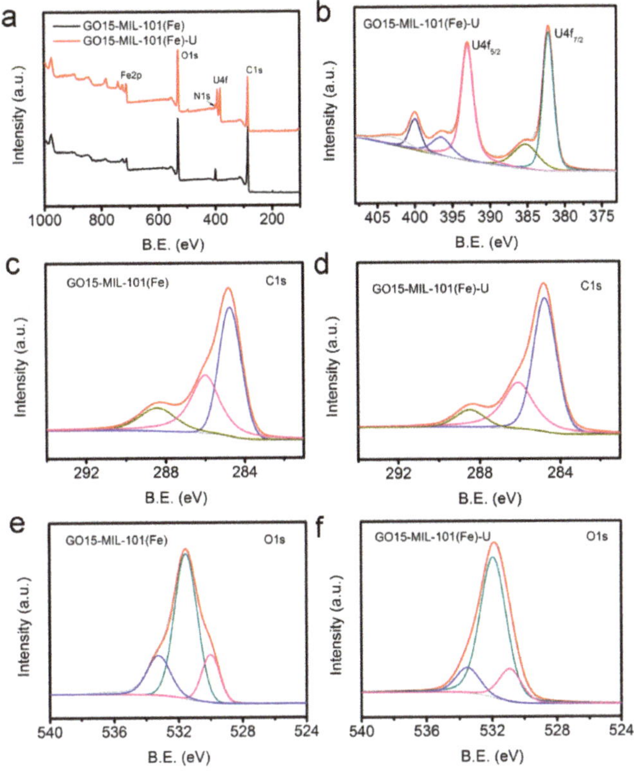

Figure 3. (**a**) The survey spectra obtained from X-ray photoelectron spectroscopy (XPS) of GO15-MIL-101(Fe) before and after the adsorption of U(VI), high-resolution spectra of U4f for (**b**) GO15-MIL-101(Fe)-U, C1s for (**c**) GO15-MIL-101(Fe) and (**d**) GO15-MIL-101(Fe)-U, O1s for (**e**) GO15-MIL-101(Fe) and (**f**) GO15-MIL-101(Fe)-U.

3.2. U(VI) Adsorption on the GO-MIL-101(Fe)

Figure 4a showed the change of the adsorption efficiency of MIL-101(Fe) and GO-MIL-101(Fe) towards U(VI) as a function of pH. With pH varies ranging from 2.0 to 11.0, the adsorption percentage of uranyl ions increased slowly firstly (Stage I), followed by a rapid improvement (Stage II) and then it was maintained at a high level (Stage III). GO-MIL-101(Fe) composite materials exhibited decreased adsorption efficiency after Stage III. However, the pH turning point of Stage I–II and Stage II–III is different for MOF and GO-MOF composites. Pure MIL-101(Fe) and GO10-MIL-101(Fe) ended stage I at pH = 4, while that for GO15-MIL-101(Fe) and GO20-MIL-101(Fe) is pH = 3. Furthermore, pure MIL-101(Fe) begins stage III at pH > 7.4. In comparison, GO-MIL-101(Fe) composite reached high level of adsorption efficiency at lower pH (GO10-MIL-101(Fe)/92.2% at pH = 7, GO15-MIL-101(Fe)/98.1% at pH = 6.5, GO20-MIL-101(Fe)/96.7% at pH = 7). Considering the complexity of contaminated water sources, the adsorption efficiency of GO15-MIL-101(Fe) was tested with different initial U concentrations. Under initial U(VI) concentration of 0.035 mg L^{-1}, 1.3 mg L^{-1} and 3 mg L^{-1}, the adsorption percentage was 40%, 93.1%, and 99.6% with 21 µg L^{-1}, 89 µg L^{-1}, and 12 µg L^{-1} U(VI) left in water (Figure S3), respectively. The lowered adsorption percentage at U(VI) concentration of 0.035 mg L^{-1} may originate from the reduced driving force. The final U concentration upon treatment of contaminated waters with C_0 = 0.035 mg L^{-1} and 3 mg L^{-1} by GO-MIL-101(Fe) can reach the requirement of the Chinese National Emission Standard (0.05 mg L^{-1}) for discharged water and the US Environmental Protection Agency (EPA) (0.03 mg L^{-1}) for potable water. This result indicated that application potential of as-prepared composite samples in the treatment of U contaminated water in a specific concentration range. Due to the highest adsorption percentage in the whole pH range, GO-15 MIL-101(Fe) was selected as the model sorbent for further adsorption experiments. Moreover, the deprotonating condition of surface functional groups can greatly affect the performance of adsorbents. Figure 4b displayed the surface charge information of the GO-MIL-101(Fe) composites and pure MIL-101(Fe). All the samples showed a decreasing trend from positive charges with pH value increased in the range of 2.0–11.0. The positive zeta potential originated from protonation of amine group in the organic linker of MIL-101(Fe), which were the dominant components of the composite samples. Compared with pure MIL-101(Fe), the composite samples had a lower zeta potential due to the introduction of GO. At each pH, GO15-MIL-101(Fe) exhibited the most negative zeta potential. The isoelectric point of GO-15MIL-101(Fe) appeared at the lowest pH. The zeta potential result of as-prepared samples agreed well with the effect of pH on the adsorption efficiency of the samples towards U(VI). For GO15-MIL-101(Fe), the surface negative charging greatly improved with the dissociation of oxygenated groups on GO nanosheets in the pH range of 3.0–7.0, which favored the adsorption of U(VI) cations. As shown in Figure S4, negative charged $UO_2(OH)_4^{2-}$ would dominate in aqueous solution with pH > 9.0 [17]. The repulsion between the $UO_2(OH)_4^{2-}$ species and the composite gave rise to the decreased adsorption efficiency at high pH value. The change of adsorption percentage with pH indicated the importance of the electrostatic interaction in the uptake of U(VI) on the composite. Considering the dispersibility of the composite and adsorption performance of adsorbents as well as the interference of complex hydrolysis product such as $UO_2(OH)_2$ precipitate at higher pH, a pH of 5.5 was used as the condition for the adsorption experiments. Furthermore, both samples presented improved adsorption efficiency towards U(VI) with the adsorbent dosage increasing (Figure 4c). The increasing trend can be attributed to increased adsorption sites introduced as the addition amount of adsorbents become larger. Further increasing the adsorbent dosage to 1 g L^{-1} gives rise to an adsorption percentage of 96%. Satisfying the treatment result can be achieved either by increasing the adsorbent addition or the adsorption steps. On the other hand, the dependence of adsorption performance on the ionic strength was investigated to eValuate the application potential of adsorbents. It is shown in Figure 4d that the amount of U(VI) adsorbed on GO-15MIL-101(Fe) fluctuated slightly in the range of ±5% as the concentration of $NaNO_3$ increased, while that of pure MIL-101(Fe) showed an overall decreasing trend. Therefore, the uptake of U(VI) on GO-15MIL-101(Fe) is controlled by inner-sphere surface complexation [8,64].

To eValuate the removal rate of uranyl ions on the MIL-101(Fe) and composite samples, the adsorption percentage of U(VI) under different times was monitored. Figure 4e showed that both GO-15-MIL-101(Fe) and MIL-101(Fe) exhibited a rapid increasing adsorption percentage of U(VI) in the first 30 min and achieved it in 60 min. In comparison with the composite sample, the adsorption efficiency of MIL-101(Fe) had much lower equilibrium adsorption efficiency (~26%). In order to further understand the adsorption process, the adsorption kinetics were simulated by a pseudo-first-order (Equation (3)) [65] and the pseudo-second-order models (Equation (4)) [66]:

Pseudo-first-order model:

$$\ln(q_e - q_t) = \ln q_e - k_1 t \tag{3}$$

Pseudo-second-order model:

$$\frac{t}{q_t} = \frac{1}{k_2 q_e^2} + \frac{t}{q_e} \tag{4}$$

In Equations (3) and (4), q_t and q_e are the adsorption amount at time t and equilibrium, respectively. k_1 and k_2 correspond to the rate constants for the pseudo-first-order model and pseudo-second-order model, respectively. The kinetic plots of uranyl ions on both MIL-101(Fe) and the composite sample can be better fitted by the pseudo-second-order model with the correlation coefficients (R^2) close to 1 (Table 1). This was also confirmed by the almost linear plots of t/q_t as a function of t (Figure 4f). Therefore, chemisorption is one of the rate-limiting processes in the adsorption kinetics of U(VI) on

GO-15-MIL-101(Fe). Moreover, the diffusion of U(VI) onto the surface of adsorbent also affected the adsorption kinetics. The more rapid adsorption performance of GO15-MIL-101(Fe) in comparison with MIL-101(Fe) may arise from the meso-macroporous structure and enlarged mesopore distribution, which served as an efficient ion diffusion channel.

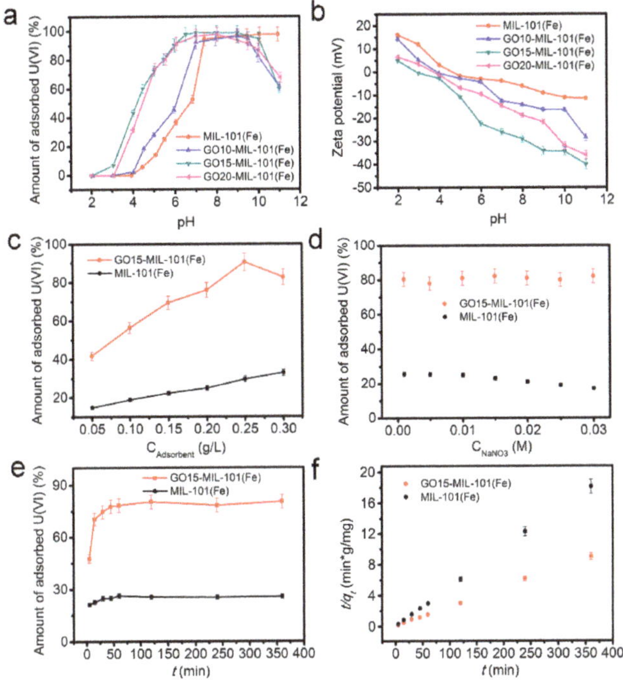

Figure 4. (**a**) The adsorption efficiency of the MIL-101(Fe) and GO-MIL-101(Fe) composite towards U(VI) as a function of pH (C_0 = 10 mg L^{-1}, T=298 K, m/V = 0.2 g L^{-1}, and I$_{NaNO_3}$ = 0.01 M), (**b**) the effect of pH on the zeta potential of the GO10 (15, 20)-MIL-101(Fe) composite and MIL-101(Fe), the effect of (**c**) adsorbent dose and (**d**) ionic strength on the adsorption efficiency of MIL-101(Fe) and GO15-MIL-101(Fe) composites. (**e**) The amount of uranyl ions adsorbed on MIL-101(Fe) and GO15-MIL-101(Fe) composites under different contact times. The experimental condition is C_0 = 10 mg L^{-1}, pH = 5.5, T = 298 K, m/V = 0.2 g L^{-1}, and I$_{NaNO_3}$ = 0.01 M for (**b–e**). (**f**) pseudo-second-order simulation result for the adsorption of the U(VI) on MIL-101(Fe) and GO15-MIL-101(Fe) composites.

Table 1. The kinetic parameters for U(VI) adsorption on MIL-101(Fe) and GO15-MIL-101(Fe).

C_0 (mg L^{-1})	Pseudo-First-Order			Pseudo-Second-Order		
	k_1 (min^{-1})	q_e (mg g^{-1})	R^2	k_2 (g mg^{-1} min^{-1})	q_e (mg g^{-1})	R^2
MIL-101(Fe)	0.361	18.40	0.577	0.0485	18.88	0.999
GO15-MIL-101(Fe)	0.179	36.93	0.961	0.0117	38.46	0.999

The real uranium-contaminated water always had several coexisting ions which could interfere with the adsorption of uranium on the adsorbents. Therefore, the adsorption percentage of uranyl ions on MIL-101(Fe) and the composite was monitored to eValuate the effect of various cations and anions (Figure 5a,b). According to the statistics shown in Table S3, GO15-MIL-101(Fe) showed an overall improved adsorption performance in comparison with MIL-101-(Fe) under different coexisted anions and cations. Especially, the adsorption of U(VI) on the composite samples was less suppressed in the presence of Ca^{2+} and CO_3^{2-}, while that for MIL-101(Fe) was largely reduced. Several studies have eVidenced that in the presence of Ca^{2+} and dissolved CO_3^{2-}, the formation of $Ca_2UO_2(CO_3)_3(aq)$ can decrease the adsorption of U(VI) on adsorbents [67]. eVen in aqueous solutions without Ca^{2+}, the existence of CO_3^{2-} with the concentration of 10 mM can greatly influence the speciation of U(VI) [68]. ($UO_2CO_3(aq)$) become dominant U(VI) species at pH = 5.5, leading to the decreased adsorption capacity. The slightly affected performance of GO15-MIL-101(Fe) indicated that the synergistic effect between GO and MIL-101(Fe) could greatly improve the capability of the adsorbents in resisting the interference of coexisted ions. The high retention of adsorption efficiency in simulated groundwater and surface water (Figure S4) indicated the potential of GO15-MIL-101(Fe) in uranium elimination from various wastewaters.

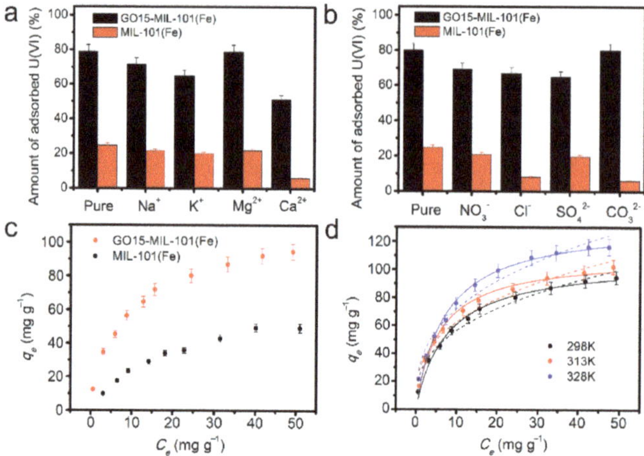

Figure 5. The effect of different (**a**) cation ions and (**b**) anion ions on the amount of adsorbed U (VI) on MIL-101(Fe) and GO15h-MIL-101(Fe) composites (C_0 = 10 mg L^{-1}, pH = 5.5, T = 298 K, m/V = 0.2 g L^{-1}, and I_{NaNO_3} = 0.01 M), (**c**) comparison of adsorption isotherms of MIL-101(Fe) and GO15-MIL-101(Fe) composites, and (**d**) adsorption isotherms of GO15-MIL-101(Fe) composite under different temperature (T = 298 K, 313 K, 328 K, pH = 5.5, m/V = 0.2 g L^{-1}, and I_{NaNO_3} = 0.01 M). The solid and dashed lines represent the Langmuir model and Freundlich model, respectively.

The maximum adsorption capacity of MIL-101(Fe) and composite samples were determined by plotting the adsorption isotherms (q_e vs. C_e) under pH = 5.5 and T = 298 K. Figure 5c indicated that with the increasing of equilibrium U(VI) concentration, the equilibrium adsorption capacity increased rapidly at first and then reached a saturated value. Obviously, the GO15-MIL-101(Fe) composite possessed much higher adsorption capacity than pure MIL-101(Fe). The adsorption isotherms were simulated through the Langmuir model (Equation (5)) and Freundlich model (Equation (6)). The Langmuir model describes monolayer adsorption onto the homogeneous surface with equivalent active sites. In contrast, the Freundlich model refers to multilayer adsorption on a surface with a heterogeneous surface.

$$q_e = \frac{bq_{max}C_e}{1+bC_e} \tag{5}$$

$$q_e = K_F C_e^{1/n_F} \tag{6}$$

where q_{max} refers to the maximum adsorption capacity, b is the Langmuir constant (L mg^{-1}), K_F and n_F represent the Freundlich constants corresponding to the capacity and intensity of adsorption, respectively. As listed in Table 2, the calculated q_{max} was 106.89 mg g^{-1} for the GO15-MIL-101(Fe) composite, which was much larger than that of MIL-101(Fe) (68.93 mg g^{-1}), respectively. The larger value of R^2 indicated that the Langmuir model fitted the adsorption isotherms of both the GO15-MIL-101(Fe) composite and MIL-101(Fe) better than the Freundlich model. This analysis confirmed the monolayer adsorption nature of the U(VI) adsorption on the samples. As summarized in Table S3, the GO15-MIL-101(Fe) composite an exhibited competitive adsorption capacity achieved from the facile one-pot synthetic process in comparison with other adsorbents.

Table 2. The simulated adsorption isotherm parameters of U(VI) on MIL-101(Fe) and GO15-MIL-101(Fe) composites at 298 K.

T (K)	Langmuir			Freundlich		
	q_{max} (mg g^{-1})	b (L mg^{-1})	R^2	K_F (mol^{1-n} L^n g^{-1})	n	R^2
MIL-101(Fe)	68.93	0.053	0.991	7.66	2.03	0.969
GO15-MIL-101(Fe)	106.89	0.132	0.989	24.96	2.824	0.973

As shown in Figure 5d, the thermodynamics of U(VI) adsorption on the GO15-MIL-101(Fe) composite was studied based on the adsorption isotherms at different temperatures. The fitted parameters based on Equations (5) and (6) are listed in Table 3. Clearly, q_{max} became larger with temperature, indicating that the adsorption of uranyl ions on GO15-MIL-101(Fe) was endothermic [69]. To eValuate how temperature affects the adsorption process, the thermodynamic parameters including the Gibbs free energy change (ΔG^0), the enthalpy change (ΔH^0), and the entropy change (ΔS^0) were achieved from the following equations:

$$\Delta G^0 = -RT \ln K^0 \tag{7}$$

$$\ln K^0 = \frac{\Delta S^0}{R} - \frac{\Delta H^0}{RT} \tag{8}$$

Firstly, K^0, the equilibrium constant, was obtained from plots of $\ln K_d$ as a function of C_e while extrapolating C_e to zero ($K_d = \frac{C_0 - C_e}{C_e} \times \frac{V}{m}$) based on the experimental data. ΔG^0 under different temperatures can be further achieved from Equation (7). Figure S5 showed a good linear relationship between $\ln K^0$ and $1/T$ for GO15-MIL-101(Fe). The slope and intercept corresponded to ΔH^0 (kJ mol^{-1}) and ΔS^0 (J mol^{-1} K^{-1}) of the adsorption, respectively. The negative ΔG^0 and its increased absolute value, as well as the positive ΔH^0, illustrated that the uptake of U(VI) by GO15-MIL-101(Fe) was spontaneous and endothermic (Table 4). This result manifested that the adsorbed heat for U(VI) dehydration was larger than the energy which was released from the process of uranyl ions binding on the surface of the adsorbents. The positive ΔS^0 (104.34 J mol^{-1} K^{-1}) and very negative ΔG^0 indicated the strong interaction between the GO15-MIL-101(Fe) and U(VI).

Table 3. The simulated adsorption isotherm parameters of U(VI) on GO15-MIL-101(Fe) composite.

T (K)	Langmuir			Freundlich		
	q_{max} (mg g^{-1})	b (L mg^{-1})	R^2	K_F (mol^{1-n} L^n g^{-1})	n	R^2
298	106.89	0.13	0.989	24.97	2.82	0.973
313	110.15	0.17	0.988	28.85	2.96	0.901
328	133.83	0.14	0.986	32.20	2.85	0.929

Table 4. The thermodynamic parameters for U(VI) adsorption on and the GO15-MIL-101(Fe) composite.

Adsorbents	ΔG^0 (kJ mol^{-1})			ΔH^0 (kJ mol^{-1})	ΔS^0 (J mol^{-1} K^{-1})
	298 K	313 K	328 K		
GO15-MIL-101(Fe)	−23.16	−24.71	−26.29	7.96	104.34

The application potential of GO15-MIL-101(Fe) was investigated through several cycles of regeneration and reusing experiments. The low adsorption efficiency of GO15-MIL-101(Fe) under pH < 3 indicated that U(VI) ions may desorb from adsorbents by decreasing the solution pH.

Therefore, nitric acid was used to desorb the uranyl ions in the regeneration experiment. As shown in Figure 6, GO15-MIL-101(Fe) showed the negligible reduction of adsorption efficiency in the first two reabsorption process. The adsorption decay became obvious after 4 cycles but gradually became steady in further tests. The performance decrease may arise from the incomplete desorption of uranyl ions and the partial aggregation of the composite sample (Figure S7). Because of the growth of MIL-101(Fe) on the surface, the aggregation did not deteriorate further, giving rise to the gradually steady reusing performance. After 8 cycles the samples still maintained ~75% of the original adsorption efficiency. The results implied that GO15-MIL-101(Fe) was a promising composite adsorbent with a good reusability.

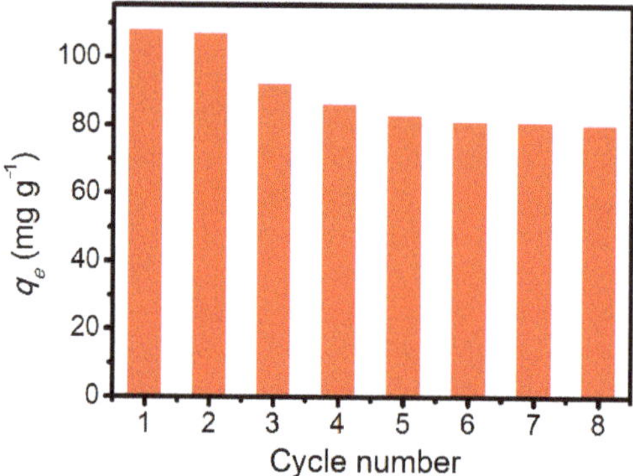

Figure 6. The recycling performance of GO15-MIL-101(Fe) in the removal of U(VI) from aqueous solution (C_0 = 10 mg L^{-1}, pH = 5.5, T = 298 K, m/V = 0.2 g L^{-1}, and I_{NaNO_3} = 0.01 M).

4. Conclusions

To conclude, the GO-MIL-101(Fe) sandwich composite was synthesized through a facile in situ growth method. By changing the amount of GO precursors, the size and distribution density of MIL-101(Fe) in the composites were easily tuned. Notably, batch adsorption experiments showed that GO-15-MIL-101(Fe) possessed superior adsorption performance with enhanced adsorption efficiency and improved adsorption capacity (106.89 mg g^{-1} at pH = 5.5, T= 298 K) compared with raw MIL-101(Fe) (68.93 mg g^{-1} at same condition). The uptake of U(VI) by GO15-MIL-101(Fe) was realized through the inner-sphere surface complexation as well as electrostatic interaction. FT-IR and XPS analysis illustrated the importance of the synergistic effect between GO nanosheets and MIL-101(Fe) octahedrons on the resultant adsorption performance. Moreover, the good reusing capability of GO15-MIL-101(Fe) indicated the application potential of such a composite material. Therefore, this work provided an effective route to design advanced composite adsorbents with satisfying radionuclide uptake ability by giving play to the functionality of individual components.

Supplementary Materials: The following are available online at http://www.mdpi.com/2076-3417/8/11/2270/s1, Table S1: Composition of groundwater and surface water used to eValuate the adsorption of U(VI) on GO15-MIL-101(Fe), Figure S1: XRD pattern of as-prepared MIL-101(Fe) and GO-MIL-101(Fe) composite, Figure S2: Nitrogen adsorption-desorption isotherm and pore size distribution of (a) and (b) MIL-101(Fe), (c) and (d) GO15-MIL-101(Fe), Table S2: Summary of surface area and pore volume obtained from N_2 adsorption isotherms, Figure S3: The adsorption efficiency of GO15-MIL-101(Fe) towards U(VI) at different initial U(VI) concentrations (0.035 mg L^{-1}, 1.3 mg L^{-1}, 3 mg L^{-1} and 10 mg L^{-1}). The inset numbers represent the adsorbent dosage, Figure S4: U(VI) speciation based on Visual MINTEQ program in the experimental adsorption solution ([U(VI)] = 10 mg L^{-1}, I = 0.01 mol L^{-1} (NaNO$_3$), and T = 25 °C), Table S3. Comparison of the adsorption efficiency of GO15-MIL-101(Fe) and MIL-101(Fe) under different coexisted ions with that in the absence of these ions, Figure S5: The adsorption efficiency of GO15-MIL-101(Fe) towards U(VI) in deionized water (pH = 5.5), simulated surface water (pH = 7.9) and simulated groundwater (pH = 8.0) at T = 298 K, and C_0 = 10 mg L^{-1}, adsorbent concentration = 0.2 g L^{-1}, Table S4: Comparison of the adsorption capacity of GO-MIL-101(Fe) composite towards U(VI) with other adsorbents, Figure S6: The plot of lnK_0 to $1/T$ of U(VI) adsorption onto GO15-MIL-101(Fe). Figure S7: The morphology of regenerated composite sample after 4 cycles, scale bar = 1 μm.

Author Contributions: B.H. conceived the project. B.H. conducted the experiment and performed the measurements with help from E.Z. and G.C. Data analysis were carried out by B.H. and E.Z. All the authors discussed the results and commented on the manuscript.

Funding: This research was funded by the National Natural Science Foundation of China (21703064) and the Fundamental Research Funds for the Central Universities (2017MS046).

Acknowledgments: The authors wish to thank the National Natural Science Foundation of China and Ministry of Education of the People's Republic of China for the funding support.

Conflicts of Interest: The authors declare no conflict of interest.

References

1. Domingo, J.L. Reproductive and developmental toxicity of natural and depleted uranium: A review. *Reprod. Toxicol.* **2001**, *15*, 603–609. [CrossRef]

2. Holdway, D.A. Uranium toxicity to two species of Australian tropical fish. *Sci. Total Environ.* **1992**, *125*, 137–158. [CrossRef]

3. Cheng, K.L.; Hogan, A.C.; Parry, D.L.; Markich, S.J.; Harford, A.J.; van Dam, R.A. Uranium toxicity and speciation during chronic exposure to the tropical freshwater fish, mogurnda mogurnda. *Chemosphere* **2010**, *79*, 547–554. [CrossRef] [PubMed]

4. Tian, G.; Geng, J.X.; Jin, Y.D.; Wang, C.L.; Li, S.Q.; Chen, Z.; Wang, H.; Zhao, Y.S.; Li, S.J. Sorption of uranium(VI) using oxime-grafted ordered mesoporous carbon cmk-5. *J. Hazard. Mater.* **2011**, *190*, 442–450. [CrossRef] [PubMed]

5. Kumar, J.R.; Kim, J.-S.; Lee, J.-Y.; Yoon, H.-S. A brief review on solvent extraction of uranium from acidic solutions. *Sep. Purif. Rev.* **2011**, *40*, 77–125. [CrossRef]

6. Manos, M.J.; Kanatzidis, M.G. Layered metal sulfides capture uranium from seawater. *J. Am. Chem. Soc.* **2012**, *134*, 16441–16446. [CrossRef] [PubMed]

7. Abney, C.W.; Mayes, R.T.; Saito, T.; Dai, S. Materials for the recovery of uranium from seawater. *Chem. Rev.* **2017**, *117*, 13935–14013. [CrossRef] [PubMed]

8. Li, Z.; Chen, F.; Yuan, L.; Liu, Y.; Zhao, Y.; Chai, Z.; Shi, W. Uranium(VI) adsorption on graphene oxide nanosheets from aqueous solutions. *Chem. Eng. J.* **2012**, *210*, 539–546. [CrossRef]

9. Hua, Y.; Wang, W.; Huang, X.; Gu, T.; Ding, D.; Ling, L.; Zhang, W.-X. Effect of bicarbonate on aging and reactivity of nanoscale zerovalent iron (nZVI) toward uranium removal. *Chemosphere* **2018**, *201*, 603–611. [CrossRef] [PubMed]

10. Shao, L.; Wang, X.; Ren, Y.; Wang, S.; Zhong, J.; Chu, M.; Tang, H.; Luo, L.; Xie, D. Facile fabrication of magnetic cucurbit [6]uril/graphene oxide composite and application for uranium removal. *Chem. Eng. J.* **2016**, *286*, 311–319. [CrossRef]

11. Sun, Y.; Ding, C.; Cheng, W.; Wang, X. Simultaneous adsorption and reduction of U(VI) on reduced graphene oxide-supported nanoscale zerovalent iron. *J. Hazard. Mater.* **2014**, *280*, 399–408. [CrossRef] [PubMed]

12. Rizescu, C.; Podolean, I.; Albero, J.; Parvulescu, V.I.; Coman, S.M.; Bucur, C.; Puche, M.; Garcia, H. N-doped graphene as a metal-free catalyst for glucose oxidation to succinic acid. *Green Chem.* **2017**, *19*, 1999–2005. [CrossRef]

13. Fei, F.; Cseri, L.; Szekely, G.; Blanford, C.F. Robust covalently cross-linked polybenzimidazole/graphene oxide membranes for high-flux organic solvent nanofiltration. *ACS Appl. Mater. Interfaces* **2018**, *10*, 16140–16147. [CrossRef] [PubMed]

14. Tan, L.; Liu, Q.; Jing, X.; Liu, J.; Song, D.; Hu, S.; Liu, L.; Wang, J. Removal of uranium(VI) ions from aqueous solution by magnetic cobalt ferrite/multiwalled carbon nanotubes composites. *Chem. Eng. J.* **2015**, *273*, 307–315. [CrossRef]

15. Fasfous, I.I.; Dawoud, J.N. Uranium (VI) sorption by multiwalled carbon nanotubes from aqueous solution. *Appl. Surf. Sci.* **2012**, *259*, 433–440. [CrossRef]

16. Luo, H.; Liu, Z.; Chao, L.; Wu, X.; Lei, X.; Chang, Z.; Sun, X. Adsorption and desorption of U(VI) on functionalized graphene oxides: A combined experimental and theoretical study. *Environ. Sci. Technol.* **2015**, *3*, 3667–3675.

17. Han, B.; Zhang, E.; Cheng, G.; Zhang, L.; Wang, D.; Wang, X. Hydrothermal carbon superstructures enriched with carboxyl groups for highly efficient uranium removal. *Chem. Eng. J.* **2018**, *338*, 734–744. [CrossRef]
18. Lv, M.; Yan, L.; Liu, C.; Su, C.; Zhou, Q.; Zhang, X.; Lan, Y.; Zheng, Y.; Lai, L.; Liu, X.; et al. Non-covalent functionalized graphene oxide (GO) adsorbent with an organic gelator for co-adsorption of dye, endocrine-disruptor, pharmaceutical and metal ion. *Chem. Eng. J.* **2018**, *349*, 791–799. [CrossRef]
19. Schierz, A.; Zänker, H. Aqueous suspensions of carbon nanotubes: Surface oxidation, colloidal stability and uranium sorption. *Environ. Pollut.* **2009**, *157*, 1088–1094. [CrossRef] [PubMed]
20. Sun, Y.B.; Shao, D.D.; Chen, C.; Yang, S.B.; Wang, X.K. Highly efficient enrichment of radionuclides on graphene oxide-supported polyaniline. *Environ. Sci. Technol.* **2013**, *47*, 9904–9910. [CrossRef] [PubMed]
21. Tan, L.; Wang, Y.; Liu, Q.; Wang, J.; Jing, X.; Liu, L.; Liu, J.; Song, D. Enhanced adsorption of uranium (VI) using a three-dimensional layered double hydroxide/graphene hybrid material. *Chem. Eng. J.* **2015**, *259*, 752–760. [CrossRef]
22. Li, Z.-J.; Wang, L.; Yuan, L.-Y.; Xiao, C.-L.; Mei, L.; Zheng, L.-R.; Zhang, J.; Yang, J.-H.; Zhao, Y.-L.; Zhu, Z.-T.; et al. Efficient removal of uranium from aqueous solution by zero-valent iron nanoparticle and its graphene composite. *J. Hazard. Mater.* **2015**, *290*, 26–33. [CrossRef] [PubMed]
23. El-Maghrabi, H.H.; Abdelmaged, S.M.; Nada, A.A.; Zahran, F.; El-Wahab, S.A.; Yahea, D.; Hussein, G.M.; Atrees, M.S. Magnetic graphene based nanocomposite for uranium scavenging. *J. Hazard. Mater.* **2017**, *322*, 370–379. [CrossRef] [PubMed]
24. Song, W.; Wang, X.; Wang, Q.; Shao, D.; Wang, X. Plasma-induced grafting of polyacrylamide on graphene oxide nanosheets for simultaneous removal of radionuclides. *Phys. Chem. Chem. Phys.* **2015**, *17*, 398–406. [CrossRef] [PubMed]
25. Hu, R.; Shao, D.; Wang, X. Graphene oxide/polypyrrole composites for highly selective enrichment of U(VI) from aqueous solutions. *Polym. Chem.* **2014**, *5*, 6207–6215. [CrossRef]
26. Linghu, W.; Yang, H.; Sun, Y.; Sheng, G.; Huang, Y. One-pot synthesis of LDH/GO composites as highly effective adsorbents for decontamination of U(VI). *ACS Sustain. Chem. Eng.* **2017**, *5*, 5608–5616. [CrossRef]
27. Li, J.-R.; Kuppler, R.J.; Zhou, H.-C. Selective gas adsorption and separation in metal-organic frameworks. *Chem. Soc. Rev.* **2009**, *38*, 1477–1504. [CrossRef] [PubMed]
28. Chen, H.; Shen, K.; Mao, Q.; Chen, J.; Li, Y. Nanoreactor of MOF-derived yolk–shell Co@C–N: Precisely controllable structure and enhanced catalytic activity. *ACS Catal.* **2018**, *8*, 1417–1426. [CrossRef]
29. Towsif Abtab, S.M.; Alezi, D.; Bhatt, P.M.; Shkurenko, A.; Belmabkhout, Y.; Aggarwal, H.; Weseliński, Ł.J.; Alsadun, N.; Samin, U.; Hedhili, M.N.; et al. Reticular chemistry in action: A hydrolytically stable MOF capturing twice its weight in adsorbed water. *Chem* **2018**, *4*, 94–105. [CrossRef]
30. Tan, L.; Wang, Y.; Liu, Q.; Wang, J.; Jing, X.; Liu, L.; Liu, J.; Song, D. Superior removal of arsenic from water with zirconium metal-organic framework UiO-66. *Sci. Rep.* **2015**, *259*, 752–760.
31. Ma, S.L.; Huang, L.; Ma, L.J.; Shim, Y.; Islam, S.M.; Wang, P.L.; Zhao, L.D.; Wang, S.C.; Sun, G.B.; Yang, X.J.; et al. Functionalized metal-organic framework as a new platform for efficient and selective removal of cadmium (II) from aqueous solution. *J. Mater. Chem. A* **2015**, *137*, 3670–3677.
32. Karmakar, S.; Dechnik, J.; Janiak, C.; De, S. Aluminium fumarate metal-organic framework: A super adsorbent for fluoride from water. *J. Hazard. Mater.* **2016**, *303*, 10–20. [CrossRef] [PubMed]
33. Campbell, J.; Burgal, J.D.S.; Szekely, G.; Davies, R.P.; Braddock, D.C.; Livingston, A. Hybrid polymer/MOF membranes for organic solvent nanofiltration (OSN): Chemical modification and the quest for perfection. *J. Membr. Sci.* **2016**, *503*, 166–176. [CrossRef]
34. Carboni, M.; Abney, C.W.; Liu, S.; Lin, W. Highly porous and stable metal-organic frameworks for uranium extraction. *Chem. Sci.* **2013**, *4*, 2396–2402. [CrossRef]
35. Kumar, P.; Pournara, A.; Kim, K.-H.; Bansal, V.; Rapti, S.; Manos, M.J. Metal-organic frameworks: Challenges and opportunities for ion-exchange/sorption applications. *Prog. Mater. Sci.* **2017**, *86*, 25–74. [CrossRef]
36. Li, L.N.; Ma, W.; Shen, S.S.; Huang, H.X.; Bai, Y.; Liu, H.W. A combined experimental and theoretical study on the extraction of uranium by amino-derived metal–organic frameworks through post-synthetic strategy. *ACS Appl. Mater. Interfaces* **2016**, *8*, 31032–31041. [CrossRef] [PubMed]
37. De Decker, J.; Folens, K.; De Clercq, J.; Meledina, M.; Van Tendeloo, G.; Du Laing, G.; Van Der Voort, P. Ship-in-a-bottle CMPO in MIL-101(Cr) for selective uranium recovery from aqueous streams through adsorption. *J. Hazard. Mater.* **2017**, *335*, 1–9. [CrossRef] [PubMed]

38. De Decker, J.; Rochette, J.; De Clercq, J.; Florek, J.; Van Der Voort, P. Carbamoylmethylphosphine oxide-functionalized MIL-101(Cr) as highly selective uranium adsorbent. *Anal. Chem.* **2017**, *89*, 5678–5682. [CrossRef] [PubMed]

39. Yang, P.; Liu, Q.; Liu, J.; Zhang, H.; Li, Z.; Li, R.; Liu, L.; Wang, J. Interfacial growth of a metal-organic framework (UiO-66) on functionalized graphene oxide (GO) as a suitable seawater adsorbent for extraction of uranium(VI). *J. Mater. Chem. A* **2017**, *5*, 17933–17942. [CrossRef]

40. Bai, Z.Q.; Yuan, L.Y.; Zhu, L.; Liu, Z.R.; Chu, S.Q.; Zheng, L.R.; Zhang, J.; Chai, Z.F.; Shi, W.Q. Introduction of amino groups into acid-resistant MOFs for enhanced U(VI) sorption. *J. Mater. Chem. A* **2015**, *3*, 525–534. [CrossRef]

41. Jabbari, V.; Veleta, J.M.; Zarei-Chaleshtori, M.; Gardea-Torresdey, J.; Villagrán, D. Green synthesis of magnetic MOF@GO and MOF@CNT hybrid nanocomposites with high adsorption capacity towards organic pollutants. *Chem. Eng. J.* **2016**, *304*, 774–783. [CrossRef]

42. Petit, C.; Bandosz, T.J. MOF–graphite oxide nanocomposites: Surface characterization and eValuation as adsorbents of ammonia. *J. Mater. Chem.* **2009**, *19*, 6521–6528. [CrossRef]

43. Hummers, W.S.; Offeman, R.E. Preparation of graphitic oxide. *J. Am. Chem. Soc.* **1958**, *80*, 1339. [CrossRef]

44. Karel, F.; Karen, L.; Ricci, N.N.; Maria, M.; Stuart, T.; Gustaaf, V.T.; Du, L.G.; Pascal, V.D.V. Fe$_3$O$_4$@MIL-101—A selective and regenerable adsorbent for the removal of as species from water. *Eur. J. Inorg. Chem.* **2016**, *2016*, 4395–4401.

45. Jiang, D.; Burrows, A.D.; Edler, K.J. Size-controlled synthesis of MIL-101(Cr) nanoparticles with enhanced selectivity for CO$_2$ over N$_2$. *CrystEngComm* **2011**, *13*, 6916–6919. [CrossRef]

46. Konkena, B.; Vasudevan, S. Understanding aqueous dispersibility of graphene oxide and reduced graphene oxide through pKa measurements. *J. Phys. Chem. Lett.* **2012**, *3*, 867–872. [CrossRef] [PubMed]

47. Zhou, X.; Huang, W.; Shi, J.; Zhao, Z.; Xia, Q.; Li, Y.; Wang, H.; Li, Z. A novel mof/graphene oxide composite GrO@MIL-101 with high adsorption capacity for acetone. *J. Mater. Chem. A* **2014**, *2*, 4722–4730. [CrossRef]

48. Saikia, M.; Saikia, L. Sulfonic acid-functionalized MIL-101(Cr) as a highly efficient heterogeneous catalyst for one-pot synthesis of 2-amino-4H-chromenes in aqueous medium. *RSC Adv.* **2016**, *6*, 15846–15853. [CrossRef]

49. Gao, L.; Li, C.-Y.V.; Yung, H.; Chan, K.-Y. A functionalized MIL-101(Cr) metal-organic framework for enhanced hydrogen release from ammonia borane at low temperature. *Chem. Commun.* **2013**, *49*, 10629–10631. [CrossRef] [PubMed]

50. Sun, X.; Xia, Q.; Zhao, Z.; Li, Y.; Li, Z. Synthesis and adsorption performance of MIL-101(Cr)/graphite oxide composites with high capacities of n-hexane. *Chem. Eng. J.* **2014**, *239*, 226–232. [CrossRef]

51. Li, X.; Guo, W.; Liu, Z.; Wang, R.; Liu, H. Quinone-modified NH$_2$-MIL-101(Fe) composite as a redox mediator for improved degradation of bisphenol A. *J. Hazard. Mater.* **2017**, *324*, 665–672. [CrossRef] [PubMed]

52. Liu, Q.; Ning, L.; Zheng, S.; Tao, M.; Shi, Y.; He, Y. Adsorption of carbon dioxide by MIL-101(Cr): Regeneration conditions and influence of flue gas contaminants. *Sci. Rep.* **2013**, *3*, 2916. [CrossRef] [PubMed]

53. Zhang, S.; Shu, X.; Zhou, Y.; Huang, L.; Hua, D. Highly efficient removal of uranium (VI) from aqueous solutions using poly(acrylic acid)-functionalized microspheres. *Chem. Eng. J.* **2014**, *253*, 55–62. [CrossRef]

54. Yang, S.-T.; Chen, S.; Chang, Y.; Cao, A.; Liu, Y.; Wang, H. Removal of methylene blue from aqueous solution by graphene oxide. *J. Colloid Interface Sci.* **2011**, *359*, 24–29. [CrossRef] [PubMed]

55. Wager, C.D.; Riggs, W.H.; Davis, L.E.; Moulder, J.F.; Meilenberg, G.E. *Handbook of X-ray Photoelectron Spectroscopy*; Perkin-Elmer Coorperation: Eden Prairie, MN, USA, 1979.

56. *VG Scientific Auger Handbook*; VG Scientific Limited: West Sussex, UK, 1989.

57. Makarova, L.G.; Shabanova, I.N.; Kodolov, V.I.; Besogonov, Y.V. X-ray photoelectron spectroscopy as a method to control the received metal–carbon nanostructures. *J. Electron Spectrosc. Relat. Phenom.* **2004**, *137–140*, 239–242. [CrossRef]

58. Lu, H.B.; Campbell, C.T.; Graham, D.J.; Ratner, B.D. Surface characterization of hydroxyapatite and related calcium phosphates by XPS and tof-sims. *Anal. Chem.* **2000**, *72*, 2886–2894. [CrossRef] [PubMed]

59. Dhankhar, S.; Bhalerao, G.; Ganesamoorthy, S.; Baskar, K.; Singh, S. Growth and comparison of single crystals and polycrystalline brownmillerite Ca$_2$Fe$_2$O$_5$. *J. Cryst. Growth* **2017**, *468*, 311–315. [CrossRef]

60. Terzyk, A.P. The influence of activated carbon surface chemical composition on the adsorption of acetaminophen (paracetamol) in vitro: Part II. TG, FTIR, and XPS analysis of carbons and the temperature dependence of adsorption kinetics at the neutral pH. *Colloids Surf. A* **2001**, *177*, 23–45. [CrossRef]

61. Liu, Q.; Guo, B.; Rao, Z.; Zhang, B.; Gong, J.R. Strong two-photon-induced fluorescence from photostable, biocompatible nitrogen-doped graphene quantum dots for cellular and deep-tissue imaging. *Nano Lett.* **2013**, *13*, 2436–2441. [CrossRef] [PubMed]

62. Han, B.; Cheng, G.; Zhang, E.; Zhang, L.; Wang, X. Three dimensional hierarchically porous ZIF-8 derived carbon/LDH core-shell composite for high performance supercapacitors. *Electrochim. Acta* **2018**, *263*, 391–399. [CrossRef]

63. Seredych, M.; Petit, C.; Tamashausky, A.V.; Bandosz, T.J. Role of graphite precursor in the performance of graphite oxides as ammonia adsorbents. *Carbon* **2009**, *47*, 445–456. [CrossRef]

64. Hayes, K.F.; Papelis, C.; Leckie, J.O. Modeling ionic strength effects on anion adsorption at hydrous oxide/solution interfaces. *J. Colloid Interface Sci.* **1988**, *125*, 717–726. [CrossRef]

65. Azizian, S. Kinetic models of sorption: A theoretical analysis. *J. Colloid Interface Sci.* **2004**, *276*, 47–52. [CrossRef] [PubMed]

66. Miyake, Y.; Ishida, H.; Tanaka, S.; Kolev, S.D. Theoretical analysis of the pseudo-second order kinetic model of adsorption. Application to the adsorption of Ag(I) to mesoporous silica microspheres functionalized with thiol groups. *Chem. Eng. J.* **2013**, *218*, 350–357. [CrossRef]

67. Dong, W.; Ball, W.P.; Liu, C.; Wang, Z.; Stone, A.T.; Bai, J.; Zachara, J.M. Influence of calcite and dissolved calcium on uranium(VI) sorption to a hanford subsurface sediment. *Environ. Sci. Technol.* **2005**, *39*, 7949–7955. [CrossRef] [PubMed]

68. Wen, T.; Wang, X.; Wang, J.; Chen, Z.; Li, J.; Hu, J.; Hayat, T.; Alsaedi, A.; Grambow, B.; Wang, X. A strategically designed porous magnetic N-doped Fe/Fe$_3$C@C matrix and its highly efficient uranium(VI) remediation. *Inorg. Chem. Front.* **2016**, *3*, 1227–1235. [CrossRef]

69. Ding, M.; Shi, W.; Guo, L.; Leong, Z.Y.; Baji, A.; Yang, H.Y. Rational design and synthesis of monodispersed hierarchical SiO$_2$@layered double hydroxide nanocomposites for efficient removal of pollutants from aqueous solution. *Chem. Eng. J.* **2017**, *5*, 6113–6121.

 *applied sciences*

Article

Tribological Capabilities of Graphene and Titanium Dioxide Nano Additives in Solid and Liquid Base Lubricants

Jankhan Patel and Amirkianoosh Kiani *

Silicon Hall: Micro/Nano Manufacturing Facility, Faculty of Engineering and Applied Science, Ontario Tech. University (UOIT), Oshawa, ON L1G 0C5, Canada; jankhankumarnayankumar.patel@uoit.net
* Correspondence: amirkianoosh.kiani@uoit.ca

Received: 5 March 2019; Accepted: 15 April 2019; Published: 19 April 2019

Abstract: In this study, the tribological behavior of both liquid (oil) and semi-liquid (grease) lubricants enhanced by multilayer graphene nano platelets and titanium dioxide nano powder was evaluated using ball-on-disk and shaft-on-plate tribo-meters. Oil samples for both 2D graphene nano platelets (GNP) and titanium nanopowders (TiNP) were prepared at three concentrations of 0.01 %w/w, 0.05 %w/w and 0.1 %w/w. In addition, 0.05% w/w mixtures of GNP and TiNP were prepared with three different ratios to analyze collective effects of both nano additives on friction and wear properties. For semi-liquid lubricants, 0.5% w/w concentrations were prepared for both nano additives for shaft-on-plate tests. Viscosity and oxidation stability tests were conducted on the liquid-base lubricants. Nano powders of both additive and substrate were analyzed using transmission electron microscopy (TEM) and scanning electron microscopy (SEM). In addition, Raman spectroscopy was conducted to characterize the graphene and titanium dioxide. The study shows that adding graphene and titanium dioxide individually sacrifices either the wear or friction of lubricants. However, use of both additives together can enhance friction resistance and wear preventive properties of a liquid lubricant significantly. For a semi-liquid lubricant, the use of both additives together and individually reduces friction compared to base grease.

Keywords: titanium dioxide; graphene; grease; base oil; friction; wear

---

1. Introduction

Increasing demand for resources gives us an opportunity to concentrate on sustainable development, and this is not possible without considering all the aspects associated with a system. For automobile and mechanical industries where mating surfaces are used to transfer maximum energy, it is important to reduce both wear and friction, which results in less energy consumption and longer functional life for moving mechanical components in machinery [1]. This can be achieved by making surfaces frictionless or by developing better lubricants that do not just control friction and wear but reduce clotting, oxidation, foaming, and corrosion problems in hydraulic systems [2].

Advancement in nano technology allows researchers to implement different manufacturing processes and evaluate applications of nano materials in a variety of fields. In the past researchers have tried nano additives such as metal [3], metal oxide and sulfide [4,5] organic–inorganic material [6], carbon based materials [7–9], and ionic materials [10] in different lubricants. These additives have shown positive effects on lubrication characteristics due to high surface area, higher diffusion rate, and small particle size. A higher diffusion rate allows lubricants to reduce micro friction, which leads to reduced wear and heat concentration in machining processes [11].

Titanium and its oxides from macro to nano scale are widely used by different industries because of their high strength, corrosion resistant nature, bio compatibility, and cost effectiveness [12,13]. For example, titanium powder is used by the marine, aeronautical, chemical, paint, and machinery industries to prevent rust and improve strength without adding more weight to the structure [12,14]. Sunscreen lotion contains a high amount of titanium dioxide to protect the skin from UVB and UVA2 rays [14,15]. However, titanium has a bad reputation in terms of sliding tribology, due to unstable friction and wear, and higher chances of seizing [12,13]. As per stability tests and microscopy findings, 36% of nano particles should be smaller than 100 nm for high dispersibility in the water [15]. In addition, studies show that TiNP produces thick and rutile film on the substrate, which can be used for rustproofing applications in tribology [16].

Graphene allotropes are joined in hexagonal shapes, which are basic structures to form carbon based materials. Carbon based materials are synthesized in versatile structures with different physical and chemical properties; thus, they have the potential to be used for different applications such as energy storage, sensors, cancer phototherapy, and electromechanical devices [17–20]. Graphene can be produced by different methods such as chemical reduction and thermal treatment by Hummer's method [21]. However, in all these methods, there is a possibility of damaging the lattice of the carbon. Also, using toxic oxidizers can result in negative effects on the environment and increase fabrication costs. Finally, greater lattice defects from graphene can have adverse effects on tribology improvement [22]. Studies have proven that graphene multilayer nano additives can be used in engine oil, bentone grease, and hydraulic applications to improve tribological and thermal properties [23–25]. However, graphene produced using different methods has shown diverse effects as a tribo improver.

In this paper, the tribological effects of graphene nano platelets (GNP) and titanium nanopowders (TiNP) individually and together on the friction and wearability of both liquid and semi-liquid lubricants were studied. In the first phase, the liquid-base lubricants were tested at 0.01% *w/w*, 0.05% *w/w*, and 0.1% *w/w* concentrations for each additive. It was found that GNP plays an important role in controlling the coefficient of friction (COF); however, adding TiNP results in lower wear under test conditions. Graphene and titanium dioxide were then added together in different ratios at a total concentration of 0.05% *w/w* to understand the effects of the two combined additives on liquid base tribology. For semi-liquid lubricant samples, nano additives were used at a high concentration of 0.5% *w/w* for each individual nano powder. Both additives were mixed together in the same way for the liquid lubrication samples. Along with tribological characterization, physical properties such as viscosity and oxidation stability were tested for liquid lubricants. Our results from tribology tests including pin-on-disk and plate-on-shaft tests show that combining both GNP and TiNP can be useful for controlling both COF and wear. The results are based on test conditions, however, the concept can be used for synthetic forms of lubricants with enhanced wearability and controlled friction according to the application requirements.

2. Experimental Methods

2.1. Materials and Sample Preparations

2.1.1. Synthesis of Graphene and Titanium Dioxide

Two-dimensional (2D) graphene nano-platelets (GNP) were synthesized from aqueous graphite in the presence of lysozyme [26]. To prepare graphene, graphite powder and protein were mixed and sonicated for 6 h to get a higher dispersibility. End product powder was collected and centrifuged to remove un-exfoliated graphite and collect smaller sheets of graphene. Lysozyme graphene is pH sensitive, and this can be useful to prevent conglomeration of graphene in the base oil. Titanium dioxide nano powder was obtained from Sigma-Aldrich. Transmission electron microscopy (TEM) revealed the size of the nano particles to be around 21 nm. The white powder of titanium dioxide had a surface area of 35–65 m^2/g. Trace metal analysis of the nano powder discloses 99.5% purity.

2.1.2. Nano Lubrication Preparation

Pure oil provided by Petro Canada Inc. was used to prepare the liquid base lubrication samples. Crystal clear oil has 99.9% purity (Product code: N200MHT). The pour point and flash point (method: Cleveland open cup) of the base oil was found to be −18 °C and 240 °C. The density of base oil was 0.865 kg/L. The GNP and TiNP were mixed at three different concentrations of 0.01%, 0.05%, and 0.1% into the base oil using a mechanical mixer at 1500 rpm for 30 min for homogeneous dispersion. TiNP and GNP were added in three ratios at 0.05% *w/w* concentration as shown in Table 1. The mechanical mixer was cleaned after each use. Samples were ultrasonicated for 30 min at room temperature to break the conglomeration of nano platelets.

Table 1. Sample characteristics of liquid base lubricants (liquid nano lubricants).

Characterization	Weight Concentration	Type of Additive in a Base Oil (99.9% Pure)
S-1	0.01% *w/w*	Graphene
S-2	0.05% *w/w*	Graphene
S-3	0.1% *w/w*	Graphene
S-4	0.01% *w/w*	Titanium dioxide
S-5	0.05% *w/w*	Titanium dioxide
S-6	0.1% *w/w*	Titanium dioxide
S-7	0.05% *w/w*	Graphene (50%) + Titanium dioxide (50%)
S-8	0.05% *w/w*	Graphene (25%) + Titanium dioxide (75%)
S-9	0.05% *w/w*	Graphene (75%) + Titanium dioxide (25%)

The semi-liquid lubricant (grease) used in this study was commercially available white grease (National Lubricating Grease Institute (NLGI) grade-2), which can be used for applications such as automotive industries and marine and farm equipment. The temperature capability of this grease is between −15 °C and 120 °C. Grease samples were rigorously prepared at 0.5% concentration as per literature review using a magnetic mixer and were also ultrasonicated for 30 min at 120 °C (Table 2).

Table 2. Sample characterization of semi-liquid lubricants (grease nano lubricants).

Characterization	Weight Concentration	Type of Additive in Grease (Commercial Grease)
G-1	0.5% *w/w*	Graphene
G-2	0.5% *w/w*	Titanium dioxide
G-3	0.5% *w/w*	Graphene (50%) + Titanium dioxide (50%)
G-4	0.5% *w/w*	Graphene (25%) + Titanium dioxide (75%)
G-5	0.5% *w/w*	Graphene (75%) + Titanium dioxide (25%)

2.2. Characterization

2.2.1. Raman Test

Nondestructive Raman spectroscopy tests were conducted using a Renishaw Raman imaging microscope System 2000 (Raman lab at UOIT, ON, Canada). To obtain the most intense excitation peaks, a 514 nm wavelength laser was used. Each test was conducted at 0.5 W power for 10 accumulations. Statistical analysis was performed to characterize the multilayer graphene and annealed titanium dioxide.

2.2.2. Viscosity

Kinematic viscosity tests were conducted as per ASTM D445 at Kinectrics, Toronto, Ontario, Canada to obtain data that can be compared to commercial oil. From the kinematic viscosity obtained at 40 °C and 100 °C, the kinematic viscosity index was calculated. For the viscosity test, the oil was kept in a calibrated glass container overnight to get it settled and afterward the change in viscosity was evaluated in a 40 °C and 100 °C warm bath.

2.2.3. Rotating Pressure Vessel Oxidation Test (RPVOT)

Anti-oxidation tests were conducted as per ASTM standard D2272 at Kinectrics to evaluate the effect of different nano additives on lubricant oxidation stability. Oxidation stability is directly proportional to the service life of the lubricant, and this can have a greater effect on the sustainability and economics of an industry. To perform the test, lubricant, water, and copper catalyst were placed in a pressurized gauge. Oxygen was then injected into the vessel at 620 kPa. The temperature was elevated to 150 °C and the vessel was rotated at 100 rpm. The total time was calculated between the pressure drop from 620 kPa to 24.5 kPa, which is noted as oxidation time in minutes.

2.2.4. TEM and SEM

Transmission electron microscopy (TEM) was conducted at The Hospital for Sick Children (Sick Kids), Toronto, Ontario, Canada to understand the morphology of the material at higher magnifications. The samples were prepared using epoxy material and sliced into 90 nm slices using a diamond knife. Morphology images were captured at Sick Kids. A detailed examination of worn disk scars was performed using scanning electron microscopy (SEM).

2.2.5. Ball-on-Disk Test

Friction and wear tests were conducted using the ball-on-disk tribometer under in situ conditions at Nanovea. To evaluate the effect of nano additives under loading conditions, a 20 N load was applied to the pin. Each test was conducted for 20 min, during which the speed of the disk was increased from 0.01 rpm to 150× rpm logarithmically. Wear track diameter was found to be 20 mm when the ball diameter was 6 mm. In addition, the test temperature was maintained at 24 °C and humidity was kept to 40% for all the tests to overcome the environmental effects on the lubricants. To calculate the wear rate, hole volume analysis was performed in situ conditions. Surface structure was analyzed as per ISO 25178 standard.

2.2.6. Shaft-on-Plate

Grease samples were studied for friction behavior with and without the presence of nano additives using the shaft-on-plate tribo tester at UOIT, Ontario, Canada. Two plates were forced from both sides of the rotary shaft to evaluate the friction coefficient at 30 N load. Friction force was obtained along with the time. Each test was conducted at room temperature for 20 min.

2.2.7. Statistical Analysis

Statistical analysis was conducted using Microsoft Excel software to maintain the authenticity of data. Moreover, 5% error was considered for TEM particle size diameter calculations. To compare and analyze the significance of various process groups, one-way ANOVA was utilized. For dimensional characterization of nano particles, 20 measurements were conducted using Image J software. All the data and results shown in this study demonstrate statistical significance unless it has been stated otherwise.

3. Results

3.1. TEM Analysis

Graphene and titanium dioxide powders tend to conglomerate; thus, when imaging the GNP and TiNP, epoxy was used to mold the powder. Nano-sliced pieces of the epoxy were observed under transmission electron microscopy (TEM) at 25.0 kx magnification of 1 μm size. Figure 1A is a TEM image of GNP, which discloses a 2D structure with multiple layers. In Figure 1B, the titanium dioxide shows round particles conglomerated in the small chains. Figure 1C indicates the mixed form of graphene and titanium dioxide, demonstrating how titanium particles stick to the graphene surfaces, which motivated us to investigate the tribological properties of both additives together.

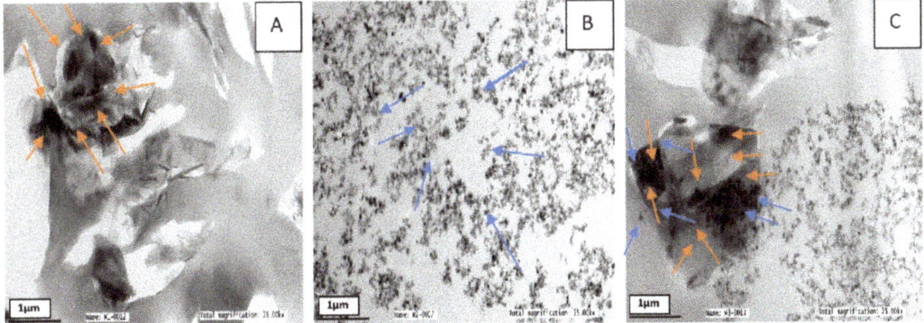

Figure 1. TEM images of (**A**) graphene, (**B**) titanium dioxide, and (**C**) graphene and titanium dioxide mixed together in powder form at 1 μm scale and 25.00 kx magnifications.

3.2. Raman Results

Raman spectroscopy allows us to determine the characteristics of the nano additives used in this research. It identifies defects in the carbon lattice by generating different intensity peaks using nondestructive methods. As illustrated in Figure 2, Raman spectroscopy illustrates the main peaks at different wavelengths of ~1354 cm^{-1} (D band), ~1581 cm^{-1} (G band), and ~2719 cm^{-1} (2D band). The sp^2 D carbon mode requires defects for activation which is illustrated by the D band intensity. In addition, the narrow G band demonstrates that optical E_{2g} phonons are in the Brillouin zone due to stretching in sp^2 chains, bond, and rings. The D + D' peak at ~2959 cm^{-1} shows the exfoliation of graphite. The intensity ratio at the D and G bands is important to understand the transformation of sp^3 to sp^2 conversion of carbons; further, it is directly proportional to the defect rate. This graphene sample, Id/Ig is 0.153, is very low compared to restored graphene, proving that this graphene has very few defects in the carbon lattice [27,28]. Along with that, lower D band intensity, with respect to G band, support less lattice defect in graphene layers [29].

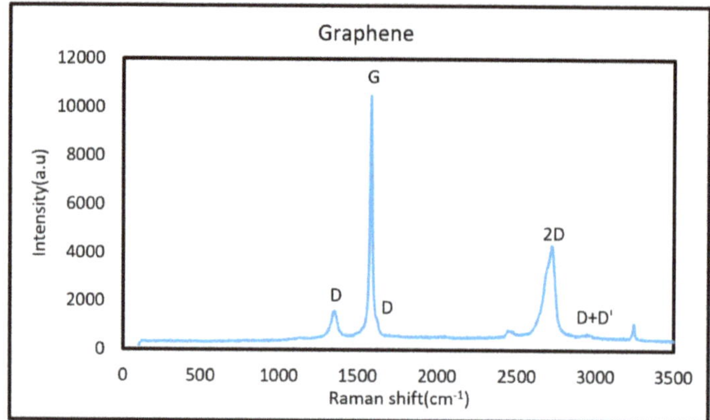

Figure 2. Raman spectroscopy of graphene nano particles.

Titanium dioxide nano powder was investigated by the Raman technique as demonstrated in Figure 3. The powder generates a primary peak at ~144 cm^{-1} and secondary intensity peaks at ~199 cm^{-1}, ~397 cm^{-1}, ~516 cm^{-1}, ~638 cm^{-1}, which is associated with different modes of energy such as E_g (strong), E_g (weak), B_{1g}, A_{1g} + B_{1g}, E_g (intermediate), respectively [30]. These peaks represent the characteristics of the anatase phase of titanium dioxide nano particles.

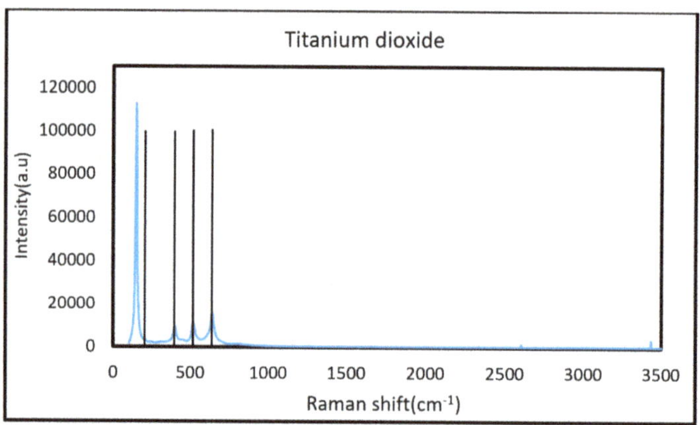

Figure 3. Raman spectroscopy of titanium dioxide nano particles.

3.3. Kinematic Viscosity

To compare the effects of different concentrations and types of nano additives in liquid base lubricants, kinematic viscosity tests were performed as per ASTM standards. Kinematic test results disclose the viscosity of nano lubricants at 40 °C and 100 °C. The resistance of lubricants to changes in temperature (viscosity index) is important to improve hydrodynamic lubrication regimes at large temperature ranges. As per the base oil data sheet provided by Petro Canada, viscosities at 40 °C and 100 °C are 41.5 mm^2/s and 6.3 mm^2/s, respectively. The calculated viscosity index for base oil is 98. However, results shown in Figure 4 prove that there is no difference in viscosity at both temperatures, nor in the viscosity index except S-3. Results demonstrate that improvement in tribology applications does not lead to sacrifices in physical properties.

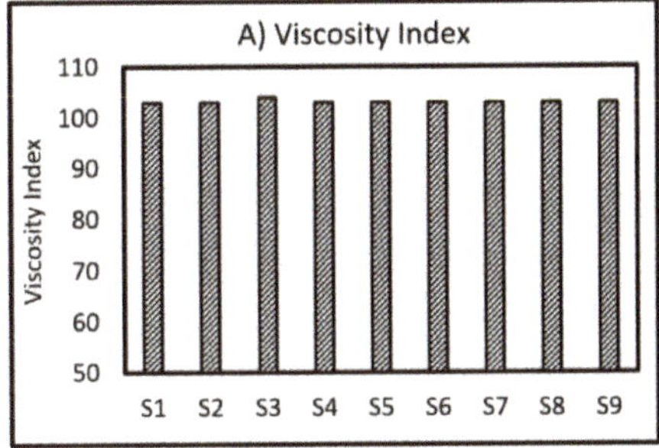
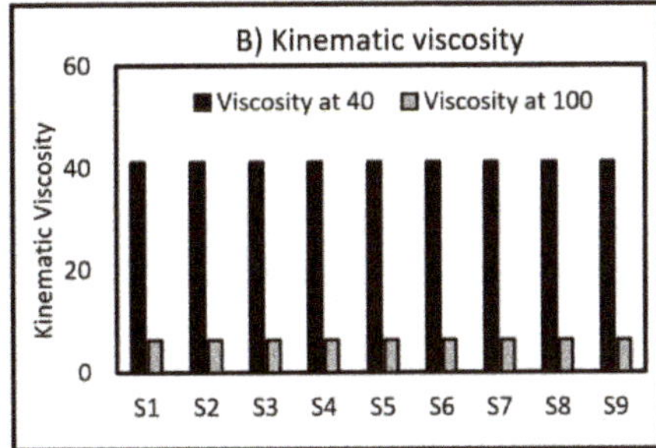

Figure 4. Viscosity index (**A**) and Kinematic viscosity (**B**) of liquid lubricants.

3.4. RPVOT

Anti-oxidation capabilities of lubricants were evaluated using the rotating pressure vessel oxidation test (RPVOT) as per ASTM standards. For sustainable improvement, it is important to consider the lubricant's service life, as lubricants that are not antioxidant can start causing cavitation in the system, which requires changing parts and lubricant. This damages the economic performance of the industry and generates environmental waste. As shown in Figure 5, there is no significant difference in oxidation time using these additives compared to base oil oxidation time. Base oil has shown oxidation time as 27.1 min under the same conditions. This proves that except for S-1, S-3, and S-6, all other samples increase oxidation stability, which allows the lubricant to have a longer service life. Test results conclude that this additive will not damage the economic performance of industries if it is added to the lubricant.

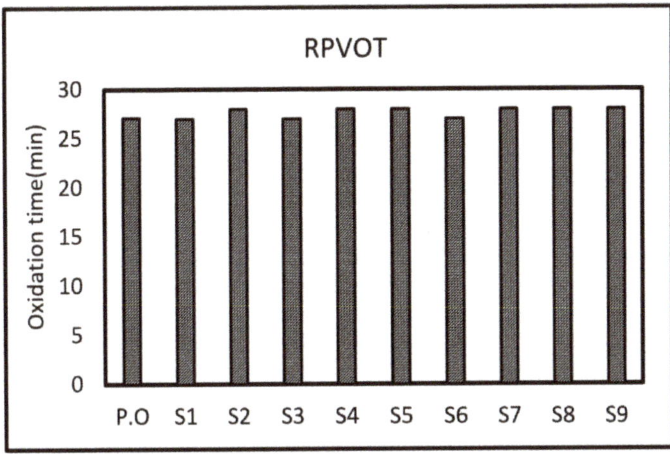

Figure 5. Rotating pressure vessel oxidation test (RPVOT) anti-oxidation test.

3.5. Friction and Wear of Liquid Base Lubricants

To obtain comparative data about the tribological effects of GNP and TiNP additives to fluid lubricants, friction and wear tests were conducted using a pin-on-disk tester. For semi-liquid samples (grease), a shaft rotatory shaft-on-plate test was used as described in the experimental setup. For liquid base lubricants, anti-scuffing and friction coefficients were calculated using in situ conditions. As demonstrated in Figure 6, a higher concentration of graphene nano additives reduces the friction coefficient; however, wear preventive properties are sacrificed. Comparing titanium dioxide with graphene illustrates that it can significantly lower the wear; however, a higher concentration of TiNP additive does not significantly affect wear preventive properties. As explained, a 0.05% concentration reveals less friction for the graphene sample. For titanium, it has lowered wear, which led us to prepare the 0.05% concentration using three different weight ratios of the additive. As illustrated, S-8 shows a lower friction coefficient and S-9 proves the lowest wear.

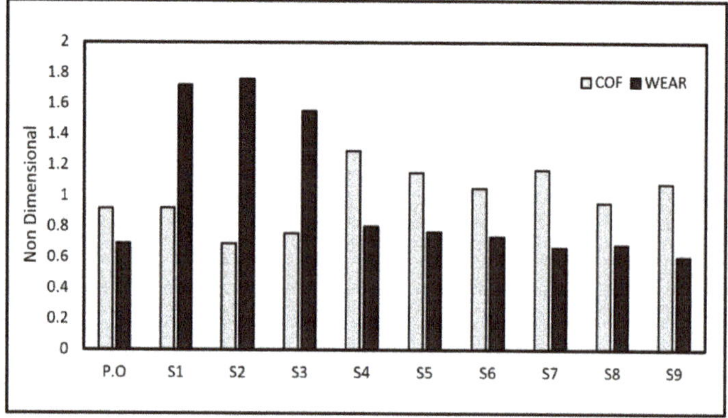

Figure 6. Friction coefficient and wear data (method for normalization: Z-score).

The results of the liquid lubricant samples enhanced by GNP and TiNP were compared separately at different concentrations. In addition, the S-2 and S-4 nano additive concentrations were found comparatively effective for hybrid nano additives of GNP and TiNP; thus, three ratios and nano additives at this concentration (0.05%) were prepared and tested. The results displayed in Figure 6 manifest that GNP has a higher wear rate compared to the coefficient of friction. However, the TiNP samples reveal a lower wear rate and higher friction coefficient. This motivated our research to evaluate the tribological effects of these additives together. S-8 illustrates moderate wear as well as friction. Along with that, S-9 shows the lowest wear but is compromised by friction. In addition, S-7 demonstrates better control on wear but it gains higher friction.

3.6. SEM Microscopy

Analysis of the worn disk helps us to understand the difference in the wear characteristics of different nano additives to lubricants. Disk wear track was evaluated at the 40 μm resolution and 1500 kx magnification in a low vacuum chamber using SEM as revealed in Figure 7. The worn surface of S-1 reveals that abrasive and adhesive wear can be seen from surface damage. This can be the result of debris penetration at the contact patch, which leads to the generation of rough surfaces. S-2 and S-3 show a higher number of pits on the surface, which can be caused by the micro crack on the substrate surface. Along with that, these pits cause an imbalance in the elastic hydrodynamic condition which can also cause corrosion on the surface [31]. Morphology of S-4 and S-6 express smooth wear tracks; however, longer surface removal marks can be seen, which indicates adhesive wear. In addition, for S-5 some plastic deformation can be found on the side of the surface. This kind of adhesive wear can be caused by micro cracks, which enlarge under continuous loading conditions. S-7 shows fatigue wear where micro cracks enlarge gradually, which causes removal of surface material that can also be called mechanical pitting. In S-8 and S-9, smooth wear track marks are visible on the substrate, which displays slight adhesive wear; however, no plastic deformation nor any abrasive wear is seen on these samples.

Further investigation of the wear track was conducted when the surface texture was examined as per ISO standard 25178. Different data were recorded including arithmetic mean height, which is important for understanding the surface waviness of the scar. As demonstrated in Figure 8, S-1 and S-5 show higher mean height with respect to the other samples, which disclose that larger peaks and valleys were found on the substrate. S-9 reveals the lowest mean height, indicating a smooth wear track.

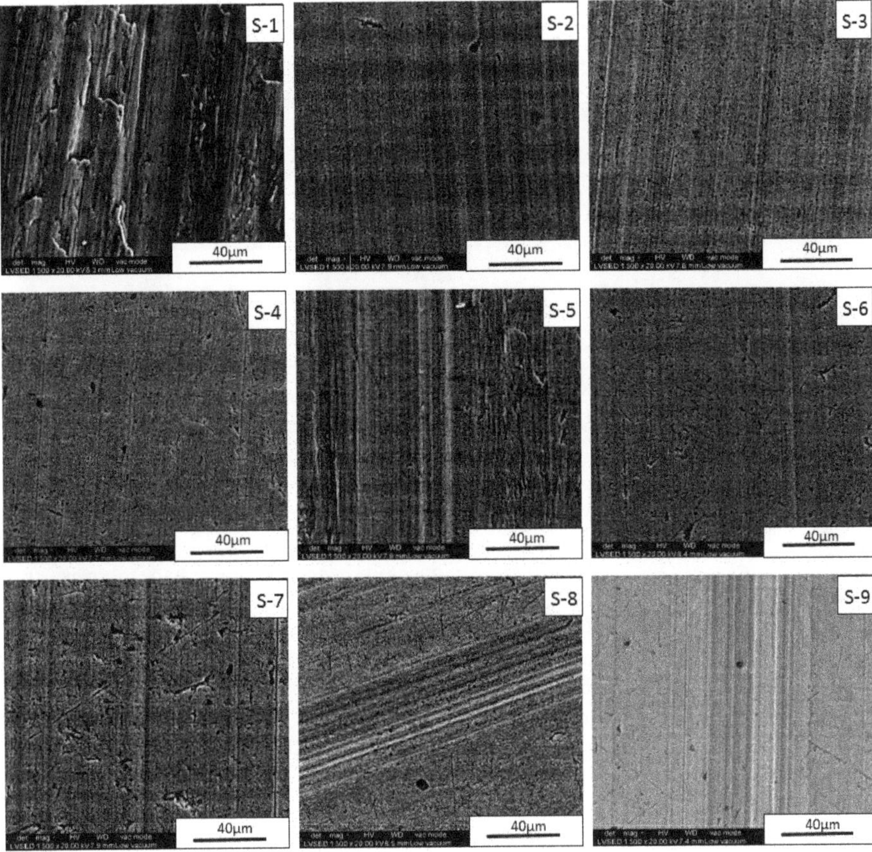

Figure 7. Worn scar evaluation in a low vacuum chamber at 40 μm scale using SEM for nine disk samples.

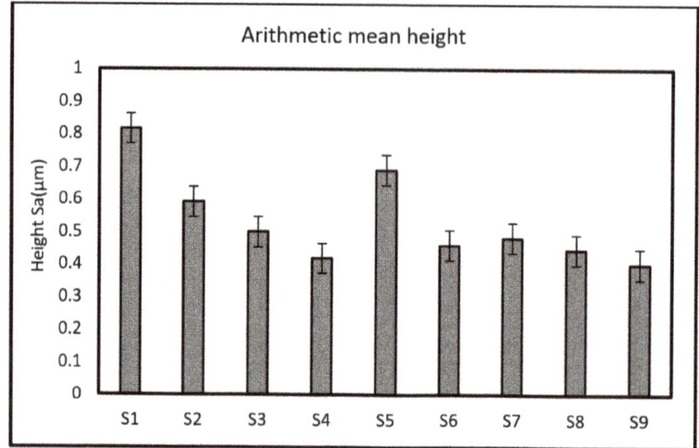

Figure 8. Arithmetic mean height of wear scar on the disk.

3.7. Frictional Results With Grease

Nano additive grease was tested using the prototyped rotatory plate-on-shaft friction system at the University of Ontario Institute of Technology. Semi-liquid lubricants (white grease) enhanced by GNP, TiNP, and a combination of both, were tested at 10 N, 20 N, and 30 N load conditions to evaluate the frictional properties of the grease. Higher concentrations of nano additives increase the dispersibility of particles in the grease. Initially, two samples were prepared where 0.5% *w/w* nano particles were added. As demonstrated in Figure 9, G1 reduces friction drastically at the 20 N load, the same as TiNP. However, at 30 N it was found that the thickness of hydrodynamic films becomes thinner, showing the minimal difference between white grease and nano particle enhanced grease. It was found that G3 (ratio of GNP/TiNP:1) manifests an increment in friction. However, G4 (lower graphene concentration and a higher titanium dioxide concentration) discloses the highest reduction in friction at the 20 N load. G5 (higher concentration of graphene and a lower concentration of titanium dioxide) shows a slight increment in average friction coefficient at the 20 N load. Each test was conducted at least 3 times and average data were considered to mitigate testing errors. To recapitulate, all 5 nano grease samples enhance the antifriction properties of the semi-liquid lubricant.

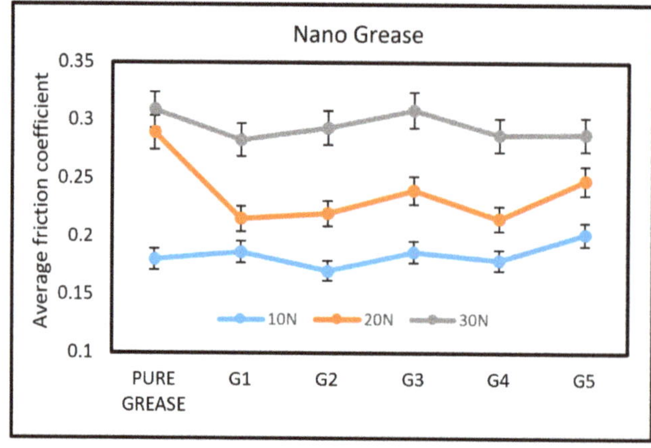

Figure 9. The average friction coefficient of nano additive grease at 10 N, 20 N, and 30 N normal load.

4. Conclusions

This study was conducted to evaluate the tribological and physical properties of graphene and titanium dioxide. Results show positive effects on base lubricants. Both nano additives showed no negative effect on oxidation life and viscosity; however, both additives reveal modifications in friction and wear properties. GNPs demonstrate a great control over friction at 0.05% *w/w* concentration; however, GNP sacrifices its wear preventive properties compared to titanium dioxide. On the other hand, titanium dioxide nano particles manifest the highest wear resistivity at 0.1% *w/w* concentration; but again, friction is higher compared to the graphene nano particles. However, for mixed nano additives of GNP and TiNP, it was found that both wear and coefficient of friction (COF) can be controlled as per application requirements, as S-9, which has a higher concentration of graphene and lower concentration of titanium dioxide, revealed the least wear. A higher concentration of titanium dioxide and a lower concentration of graphene showed more control of friction than wear.

For semi-liquid lubricants (white grease), both GNP and TiNP additives manifest lower friction coefficients (COF). In addition, the amalgamation of both nano additives lower friction compared to pure grease at the 20 N load. There is not much change in the friction coefficient at the 30 N high load because of the extremely thin film of grease on the shaft, which does not allow additives to enter the point of contact. However, at the 20 N load, a higher concentration of GNP and a lower concentration of TiNP indicates the highest reduction in friction compared to the pure grease.

Author Contributions: Conceptualization, A.K.; methodology, J.P. and A.K.; validation, A.K.; formal analysis, J.P.; investigation, J.P.; resources, A.K.; data curation, A.K.; writing—original draft preparation, J.P. and A.K.; writing—review and editing, A.K. and J.P.; visualization, J.P.; supervision, A.K.; project administration, A.K.; funding acquisition, A.K.

Funding: This research was partially funded by the National Sciences and Engineering Research Council (NSERC).

Acknowledgments: The authors would like to acknowledge research support for rotatory plate-on-shaft friction system provided by Ahmad Barari at Ontario Tech. University.

Conflicts of Interest: The authors declare no conflicts of interest.

References

1. Berman, D.; Erdemir, A.; Sumant, A.V. Graphene: A new emerging lubricant. *Mater. Today* **2014**, *17*, 31–42. [CrossRef]
2. Stachowiak, G.; Batchelor, A. Lubricants and their composition. In *Engineering Tribology*; Elsevier: Oxford, UK, 2006; pp. 51–101.
3. Choi, C.; Jung, M.; Choi, Y.; Lee, J.; Oh, J. Tribological properties of lubricating oil-based nano fluids with metal/carbon nanoparticles. *J. Nanosci. Nanotechnol.* **2011**, *11*, 368–371. [CrossRef]
4. Alves, S.; Barros, B.; Trajano, M.; Ribeiro, K.; Moura, E. Tribological behavior of vegetable oil-based lubricants with nanoparticles of oxides in boundary lubrication conditions. *Tribol. Int.* **2013**, *65*, 28–36. [CrossRef]
5. Kang, X.; Wang, B.; Zhu, L.; Zhu, H. Synthesis and tribological property study of oleic acid-modified copper sulfide nanoparticles. *Wear* **2008**, *265*, 150–154. [CrossRef]
6. Kim, D.; Archer, L.A. Nanoscale organic- inorganic hybrid lubricants. *Langmuir* **2011**, *27*, 3083–3094. [CrossRef] [PubMed]
7. Patel, J.; Kiani, A. Effects of reduced graphene oxide (rGO) at different concentrations on tribological properties of liquid base lubricants. *Lubricants* **2019**, *7*, 11. [CrossRef]
8. Kamel, B.M.; Mohamed, A.; El Sherbiny, M.; Abed, K. Tribological behavior of calcium grease containing carbon nanotubes additives. *Ind. Lubr. Tribol.* **2016**, *68*, 723–728. [CrossRef]
9. Liang, S.; Shen, Z.; Yi, M.; Liu, L.; Zhang, X.; Ma, S. In-situ exfoliated graphene for high-performance water-based lubricants. *Carbon* **2016**, *96*, 1181–1190. [CrossRef]
10. Patel, A.; Guo, H.; Iglesias, P. Study of the lubricating ability of Protic ionic liquid on an aluminum–steel contact. *Lubricants* **2018**, *6*, 66. [CrossRef]
11. Bhushan, B. Nanotribology and nanomechanics. *Wear* **2005**, *259*, 1507–1531. [CrossRef]

12. Kang, J.; Wang, M.; Yue, W.; Fu, Z.; Zhu, L.; She, D.; Wang, C. Tribological behavior of titanium alloy treated by nitriding and surface texturing composite technology. *Materials* **2019**, *12*, 301. [CrossRef] [PubMed]

13. Dong, H. Tribological properties of titanium-based alloys. In *Surface Engineering of Light Alloys*; Elsevier: Amsterdam, The Netherlands, 2010; pp. 58–80.

14. Rai, R.; Shanmuga, S.C.; Srinivas, C. Update on photo protection. *Indian J. Dermatol.* **2012**, *57*, 335. [CrossRef] [PubMed]

15. Weir, A.; Westerhoff, P.; Fabricius, L.; Hristovski, K.; von Goetz, N. Titanium dioxide nanoparticles in food and personal care products. *Environ. Sci. Technol.* **2012**, *46*, 2242–2250. [CrossRef] [PubMed]

16. Mohammed, M.T.; Khan, Z.A.; Siddiquee, A.N. Surface modifications of titanium materials for developing corrosion behavior in human body environment: A review. *Procedia Mater. Sci.* **2014**, *6*, 1610–1618. [CrossRef]

17. Hong, J.-Y.; Jang, J. Micro patterning of graphene sheets: Recent advances in techniques and applications. *J. Mater. Chem.* **2012**, *22*, 8179–8191. [CrossRef]

18. Zhang, H.; Gruener, G.; Zhao, Y. Recent advancements of graphene in biomedicine. *J. Mater. Chem. B* **2013**, *1*, 2542–2567. [CrossRef]

19. Li, P.F.; Zhou, H.; Cheng, X. Investigation of a hydrothermal reduced graphene oxide nano coating on Ti substrate and its nano-tribological behavior. *Surf. Coat. Technol.* **2014**, *254*, 298–304. [CrossRef]

20. Berman, D.; Erdemir, A.; Sumant, A.V. Reduced wear and friction enabled by graphene layers on sliding steel surfaces in dry nitrogen. *Carbon* **2013**, *59*, 167–175. [CrossRef]

21. Park, S.; Ruoff, R.S. Chemical methods for the production of graphene. *Nat. Nanotechnol.* **2009**, *4*, 217. [CrossRef]

22. Penkov, O.; Kim, H.-J.; Kim, H.-J.; Kim, D.-E. Tribology of graphene: A review. *Int. J. Precis. Eng. Manuf.* **2014**, *15*, 577–585. [CrossRef]

23. Fan, X.; Xia, Y.; Wang, L.; Li, W. Multilayer graphene as a lubricating additive in bentone grease. *Tribol. Lett.* **2014**, *55*, 455–464. [CrossRef]

24. Eswaraiah, V.; Sankaranarayanan, V.; Ramaprabhu, S. Graphene-based engine oil nano fluids for tribological applications. *ACS Appl. Mater. Interfaces* **2011**, *3*, 4221–4227. [CrossRef]

25. Zin, V.; Barison, S.; Agresti, F.; Colla, L.; Pagura, C.; Fabrizio, M. Improved tribological and thermal properties of lubricants by graphene based nano-additives. *RSC Adv.* **2016**, *6*, 59477–59486. [CrossRef]

26. Joseph, D.; Tyagi, N.; Ghimire, A.; Geckeler, K.E. A direct route towards preparing pH-sensitive graphene nanosheets with anti-cancer activity. *RSC Adv.* **2014**, *4*, 4085–4093. [CrossRef]

27. Kudin, K.N.; Ozbas, B.; Schniepp, H.C.; Prud'Homme, R.K.; Aksay, I.A.; Car, R. Raman spectra of graphite oxide and functionalized graphene sheets. *Nano Lett.* **2008**, *8*, 36–41. [CrossRef] [PubMed]

28. Wang, H.L.; Robinson, J.T.; Li, X.L.; Dai, H.J. Solve thermal reduction of chemically exfoliated graphene sheets. *J. Am. Chem. Soc.* **2009**, *131*, 9910–9911. [CrossRef] [PubMed]

29. Croce, A.; Arrais, A.; Rinaudo, C. Raman micro-spectroscopy identifies carbonaceous particles lying on the surface of crocidolite, amosite, and chrysotile fibers. *Minerals* **2018**, *8*, 249. [CrossRef]

30. Gupta, S.K.; Desai, R.; Jha, P.K.; Sahoo, S.; Kirin, D. Titanium dioxide synthesized using titanium chloride: Size effect study using raman spectroscopy and photoluminescence. *J. Raman Spectrosc.* **2010**, *41*, 350–355. [CrossRef]

31. Souza, J.C.; Henriques, M.; Teughels, W.; Ponthiaux, P.; Celis, J.-P.; Rocha, L.A. Wear and corrosion interactions on titanium in oral environment: Literature review. *J. Bio- Tribo-Corros.* **2015**, *1*, 13. [CrossRef]

MDPI

St. Alban-Anlage 66

4052 Basel

Switzerland

Tel. +41 61 683 77 34

Fax +41 61 302 89 18

www.mdpi.com

Applied Sciences Editorial Office

E-mail: applsci@mdpi.com

www.mdpi.com/journal/applsci